STUDENT WORKBOOK

VOLUME 2

college physics

a strategic approach **4e**

knight • jones • field

randall d. knight

California Polytechnic State University, San Luis Obispo

Editor in Chief, Director Physical Science Courseware Portfolio: Jeanne Zalesky
Courseware Portfolio Manager: Darien Estes
Senior Content Producer: Martha Steele
Managing Producer: Kristen Flathman
Courseware Director, Content Development: Jennifer Hart
Senior Analyst, Content Development, Science: Suzanne Olivier
Courseware Editorial Assistant: Kristen Stephens
Rich Media Content Producer: Dustin Hennessey
Full-Service Vendor: Nesbitt Graphics/Cenveo
Compositor: Nesbitt Graphics/Cenveo
Art and Design Director: Mark Ong, Side By Side Studios
Interior/Cover Designer: tani hasegawa
Cover Printer: LSC Communications
Printer: LSC Communications
Manufacturing Buyer: Stacey J. Weinberger/LSC Communications
Product Marketing Manager, Physical Sciences: Elizabeth Bell

Cover Photo Credit: MirageC/Getty

ISBN 10: 0-134-72480-1; ISBN 13: 978-0-134-72480-5
1 17
www.pearson.com

Table of Contents

Preface

It is highly unlikely that one could learn to play the piano by only reading about it. Similarly, reading physics from a textbook is not the same as doing physics. To develop your ability to do physics, your instructor will assign problems to be solved both for homework and on tests. Unfortunately, it is our experience that jumping right into problem solving after reading and hearing about physics often leads to poor "playing" techniques and an inability to solve problems for which the student has not already been shown the solution (which isn't really "solving" a problem, is it?). Because improving your ability to solve physics problems is one of the major goals of your course, time spent developing techniques that will help you do this is well spent.

Learning physics, as in learning any skill, requires regular practice of the basic techniques. That is what this *Student Workbook* is all about. The workbook consists of exercises that give you an opportunity to practice techniques and strengthen your understanding of concepts presented in the textbook and in class. These exercises are intended to be done on a daily basis, right after the topics have been discussed in class and are still fresh in your mind. Successful completion of the workbook exercises will prepare you to tackle the more quantitative end-of-chapter homework problems in the textbook.

You will find that many of the exercises are *qualitative* rather than *quantitative.* They ask you to draw pictures, interpret graphs, use ratios, write short explanations, or provide other answers that do not involve calculations. A few math-skills exercises will ask you to explore the mathematical relationships and symbols used to quantify physics concepts but do not require a calculator. The purpose of all of these exercises is to help you develop the basic thinking tools you'll later need for quantitative problem solving. It is highly recommended that you do these exercises *before* starting the end-of-chapter problems.

One example from Chapter 4 illustrates the purpose of this *Student Workbook.* In that chapter, you will read about a technique called a "free-body diagram" that is helpful for solving problems involving forces. Sometimes, students mistakenly think that the diagrams are used by the instructor only for teaching purposes and may be abandoned once Newton's laws are fully understood. On the contrary, professional physicists with decades of problem-solving experience still routinely use these diagrams to clarify the problem and set up the solution. Many of the other techniques practiced in the workbook, such as ray diagrams, graphing relationships, sketching field lines and equipotentials, etc., fall in the same category. They are used at all levels of physics, not just as a beginning exercise. And many of these techniques, such as analyzing graphs and exploring multiple representations of a situation, have important uses outside of physics. Time spent practicing these techniques will serve you well in other endeavors.

You will find that the exercises in this workbook are keyed to specific sections of the textbook in order to let you practice the new ideas introduced in that section. You should keep the text beside you as you work and refer to it often. You will usually find Tactics Boxes, figures, or examples in the textbook that are directly relevant to the exercises. When asked to draw figures or diagrams, you should attempt to draw them so that they look much like the figures and diagrams in the textbook.

Because the exercises go with specific sections in the text, you should answer them on the basis of information presented in *just* that section (and prior sections). You may have learned new ideas in Section 7 of a chapter, but you should not use those ideas when answering questions from Section 4. There will be ample opportunity in the Section 7 exercises to use that information there.

You will need a few "tools" to complete the exercises. Many of the exercises will ask you to *color code* your answers by drawing some items in black, others in red, and perhaps yet others in blue. You need to purchase a few colored pencils to do this. The authors highly recommend that you work in pencil, rather than ink, so that you can easily erase. Few are the individuals who make so few mistakes as to be able to work in ink! In addition, you'll find that a small, easily carried six-inch ruler will come in handy for drawings and graphs.

As you work your way through the textbook and this workbook, you will find that physics is a way of *thinking* about how the world works and why things happen as they do. We will primarily be interested in finding relationships, seeking explanations, and developing techniques to make use of these relationships, only secondarily in computing numerical answers. In many ways, the thinking tools developed in this workbook are what the course is all about. If you take the time to do these exercises regularly and to review the answers, in whatever form your instructor provides them, you will be well on your way to success in physics.

To the instructor: The exercises in this workbook can be used in many ways. You can have students work on some of the exercises in class as part of an active-learning strategy. Or you can do the same in recitation sections or laboratories. This approach allows you to discuss the answers immediately, to answer student questions, and to improvise follow-up exercises when needed. Having the students work in small groups (two to four students) is highly recommended.

Alternatively, the exercises can be assigned as homework. The pages are perforated for easy tear-out, and the page breaks are in logical places so that you can assign the sections of a chapter that you would likely cover in one day of class. Exercises should be assigned immediately after presenting the relevant information in class and should be due at the beginning of the next class. Collecting them at the beginning of class, and then going over two or three that are likely to cause difficulty, is an effective means of quickly reviewing major concepts from the previous class and launching a new discussion.

If used as homework, it is *essential* for students to receive *prompt* feedback. Ideally, this would occur by having the exercises graded, with written comments, and returned at the next class meeting. Posting fairly detailed answers on a course website also works. Lack of prompt feedback can negate much of the value of these exercises. Placing similar qualitative/graphical questions on quizzes and exams, and telling students at the beginning of the term that you will do so, encourages students to take the exercises seriously and to check the answers.

Student feedback from end-of-term questionnaires reveals three prevalent attitudes toward the workbook exercises:

i. They think it is an unreasonable amount of work.
ii. They agree that the assignments force them to keep up and not get behind.
iii. They recognize, by the end of the term, that the workbook is a valuable learning tool.

However you choose to use these exercises, they will significantly strengthen your students' conceptual understanding of physics.

Following the workbook exercises are optional Dynamics Worksheets, Momentum Worksheets, and Energy Worksheets for use with end-of-chapter problems in Parts I and II of the textbook. Their use is recommended to help students acquire good problem-solving habits early in the course. If you wish your students to use these, have them make enough photocopies for use throughout the term.

Answers to all workbook exercises are provided as pdf files and can be downloaded from the Instructor Resource Area in Mastering™ Physics as well as from the textbook's Instructor Resource Center at www.pearson.com.

Acknowledgments: The author would like to thank Cenveo Publishing Services for their production of the workbook.

17 Wave Optics

17.1 What is Light?

1. A light wave travels from vacuum, through a transparent material, and back to vacuum. What is the index of refraction of this material? Explain.

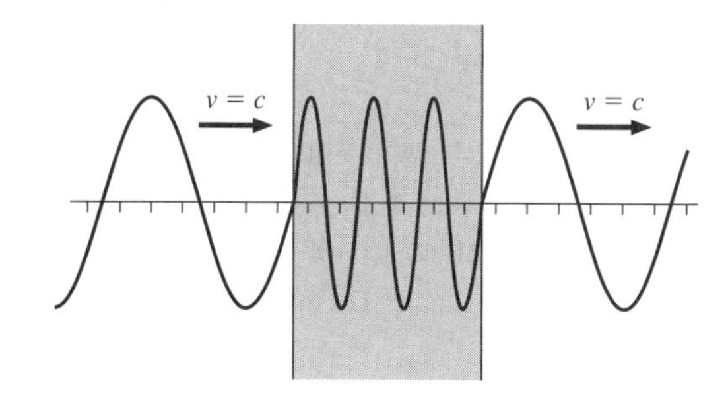

2. A light wave travels from vacuum, through a transparent material whose index of refraction is $n = 2.0$, and back to vacuum. Finish drawing the snapshot graph of the light wave at this instant.

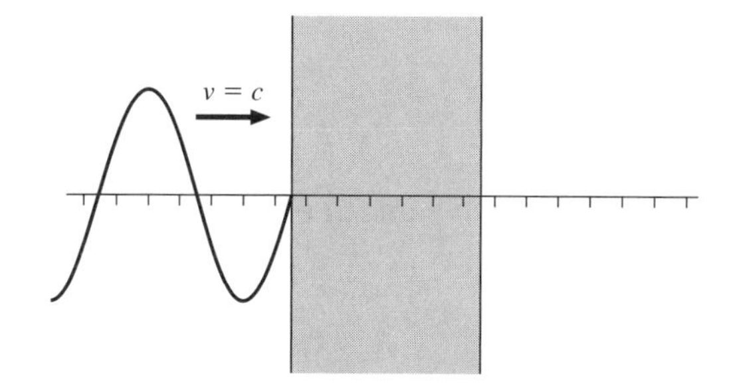

17.2 The Interference of Light

3. The figure shows the light intensity recorded by a detector in an interference experiment. Notice that the light intensity comes "full on" at the edges of each maximum, so this is *not* the intensity that would be recorded in Young's double-slit experiment.
   a. Draw a graph of light intensity versus position on the film. Your graph should have the same horizontal scale as the "photograph" above it.
   b. Is it possible to tell, from the information given, what the wavelength of the light is? If so, what is it? If not, why not?

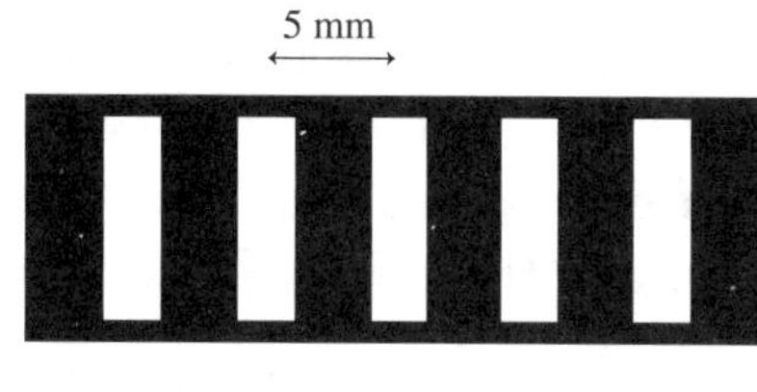

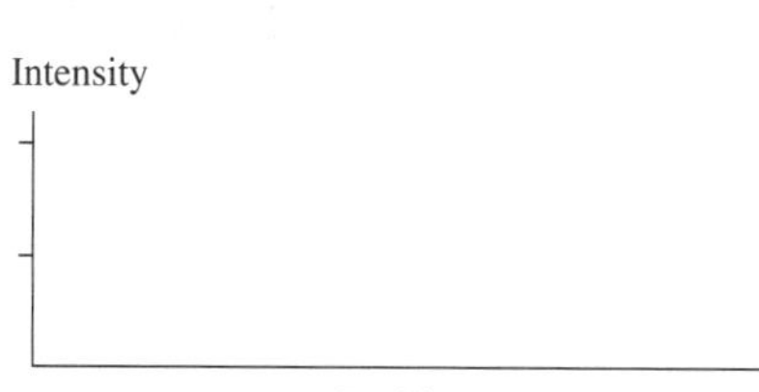

4. The graph shows the light intensity on the viewing screen during a double-slit interference experiment. Draw the "photograph" that would be recorded if a light detector were placed at the position of the screen. Your "photograph" should have the same horizontal scale as the graph above it. Be as accurate as you can. Let the white of the paper be the brightest intensity and a very heavy pencil shading be the darkest.

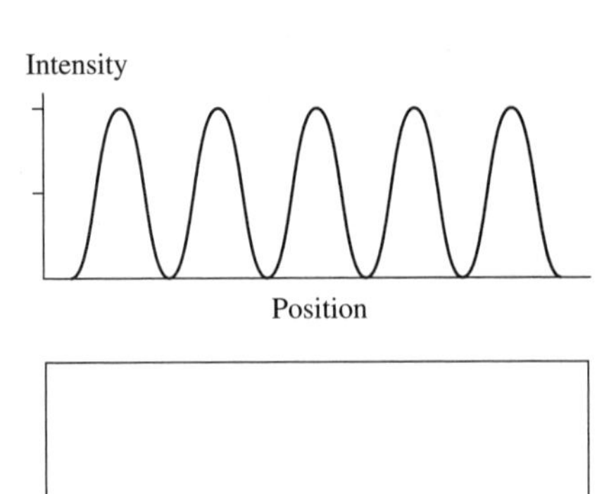

5. The figure shows the viewing screen in a double-slit experiment. For questions a–c, will the fringe spacing increase, decrease, or stay the same? Give an explanation for each.

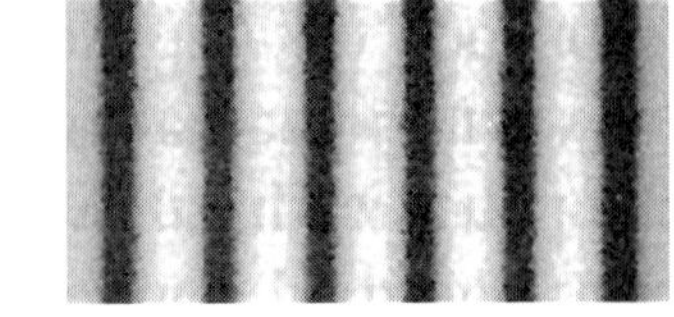

a. The distance to the screen is increased.

b. The spacing between the slits is increased.

c. The wavelength of the light is increased.

6. In a double-slit experiment, we usually see the light intensity on a viewing screen. However, we can use smoke to make the light visible as it propagates between the slits and the screen. Consider a double-slit experiment in a smoke-filled room. What kind of light and dark pattern would you see if you looked down on the experiment from above? Draw the pattern on the figure below. Shade the areas that are dark and leave the white of the paper for the bright areas. **Hint:** What is the condition for constructive interference? For destructive?

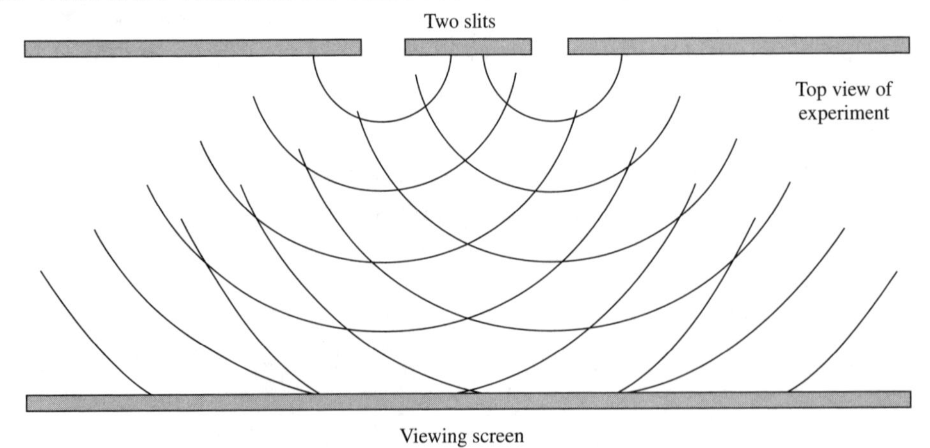

17.3 The Diffraction Grating

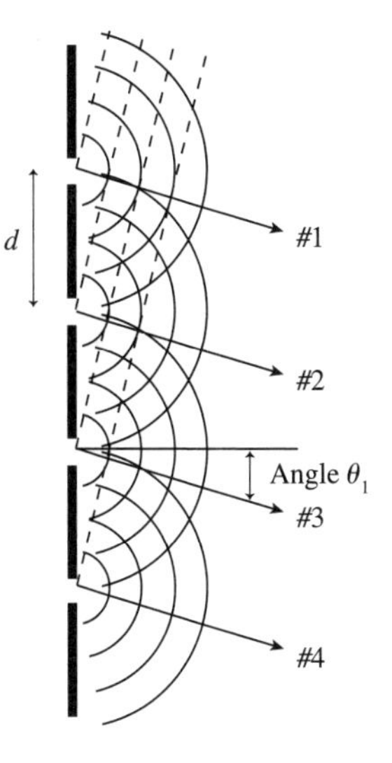

7. The figure shows four slits in a diffraction grating. A set of circular wave crests is shown spreading out from each slit. Four wave paths, numbered 1 to 4, are shown leaving the slits at angle θ_1. The dotted lines are drawn perpendicular to the paths of the waves.
   a. Use a colored pencil or heavy shading to show *on the figure* the extra distance traveled by wave 1 that is not traveled by wave 2.
   b. How many extra wavelengths does wave 1 travel compared to wave 2? Explain how you can tell from the figure.

   c. How many extra wavelengths does wave 2 travel compared to wave 3?

   d. As these four waves combine at some large distance from the grating, will they interfere constructively, destructively, or in between? Explain.

8. Suppose the wavelength of the light in Exercise 7 is doubled. (Imagine erasing every other wave front in the picture.) Would the interference at angle θ_1 then be constructive, destructive, or in between? Explain. Your explanation should be based on the figure, not on some equation.

9. Suppose the slit spacing d in Exercise 7 is doubled while the wavelength is unchanged. Would the interference at angle θ_1 then be constructive, destructive, or in between? Again, base your explanation on the figure.

10. This is the interference pattern on a viewing screen behind 2 slits. How would the pattern change if the 2 slits were replaced by 20 slits having the *same spacing d* between adjacent slits?

    a. Would the number of fringes on the screen increase, decrease, or stay the same?

    b. Would the fringe spacing increase, decrease, or stay the same?

    c. Would the width of each fringe increase, decrease, or stay the same?

    d. Would the brightness of each fringe increase, decrease, or stay the same?

17.4 Thin-Film Interference

11. The figure shows a wave transmitted from air through a thin oil film on water. The film has a thickness $t = \lambda_{oil}/2$, where λ_{oil} is the wavelength of the light while in the oil.

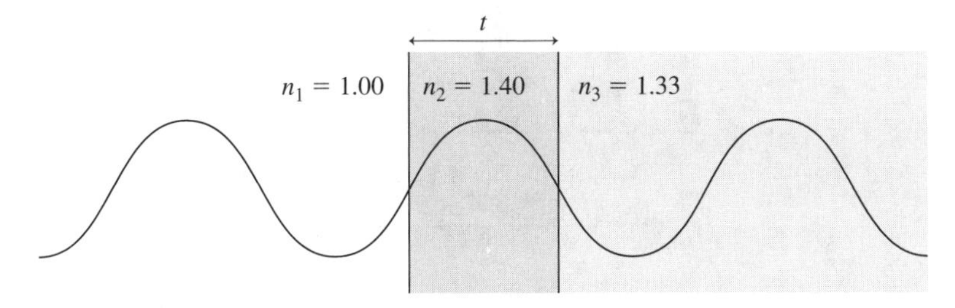

a. Referring to the indices of refraction shown on the figure, indicate at each boundary with a Y (yes) or N (no) whether the reflected wave undergoes a phase change at the boundary.

b. Notice the indices of refraction shown on the figure. At each of the two boundaries, write at the bottom of the figure a Y (yes) or N (no) to indicate whether the reflected wave undergoes a phase change at that boundary.

c. Do the two reflected waves interfere constructively, destructively, or in between? Explain.

12. The figure shows a wave transmitted from air through a thin oil film on glass. The film has a thickness $t = \lambda_{oil}/2$, where λ_{oil} is the wavelength of the light while in the oil.

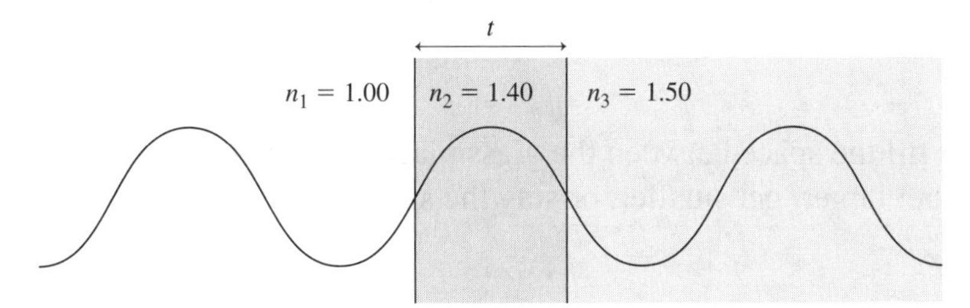

a. Referring to the indices of refraction shown on the figure, indicate at each boundary with a Y (yes) or N (no) whether the reflected wave undergoes a phase change at the boundary.

b. Notice the indices of refraction shown on the figure. At each of the two boundaries, write at the bottom of the figure a Y (yes) or N (no) to indicate whether the reflected wave undergoes a phase change at that boundary.

c. Do the two reflected waves interfere constructively, destructively, or in between? Explain.

13. The figure shows the fringes seen due to a wedge of air between two flat glass plates that touch at one end and are illuminated by light of wavelength $\lambda = 500$ nm.

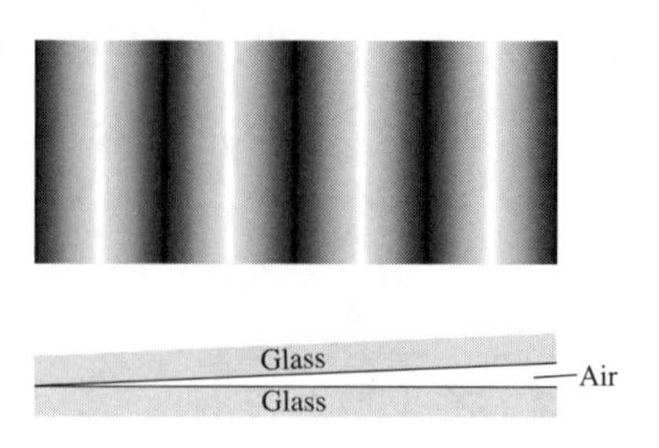

a. By how much does the wedge of air increase in thickness as you move from one dark fringe to the next dark fringe? Explain.

b. By how much does the wedge of air increase in thickness from one end of the above figure to the other?

c. Suppose you fill the space between the glass plates with water. Will the spacing between the dark fringes get larger, get smaller, or stay the same? Explain.

17.5 Single-Slit Diffraction

14. Plane waves of light are incident on two narrow, closely-spaced slits. The graph shows the light intensity seen on a screen behind the slits.

    a. Draw a graph on the bottom axes to show the light intensity on the screen if the right slit is blocked, allowing light to go only through the left slit.

    b. Explain why the graph will look this way.

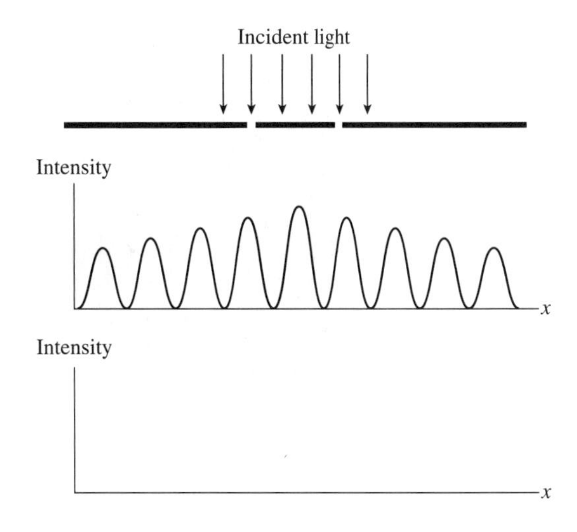

15. The graph shows the light intensity on a screen behind a 0.2-mm-wide slit illuminated by light with a 500 nm wavelength.

    a. Draw a *picture* in the box of how a photograph taken at this location would look. Use the same horizontal scale, so that your picture aligns with the graph above. Let the white of the paper represent the brightest intensity and the darkest, you can draw with a pencil or pen, be the least intensity.

    b. Using the same horizontal scale as in part a, draw graphs showing the light intensity if

       i. $\lambda = 250$ nm, $a = 0.2$ mm.

       ii. $\lambda = 1000$ nm, $a = 0.2$ mm.

       iii. $\lambda = 500$ nm, $a = 0.1$ mm.

       iv. $\lambda = 500$ nm, $a = 0.4$ mm.

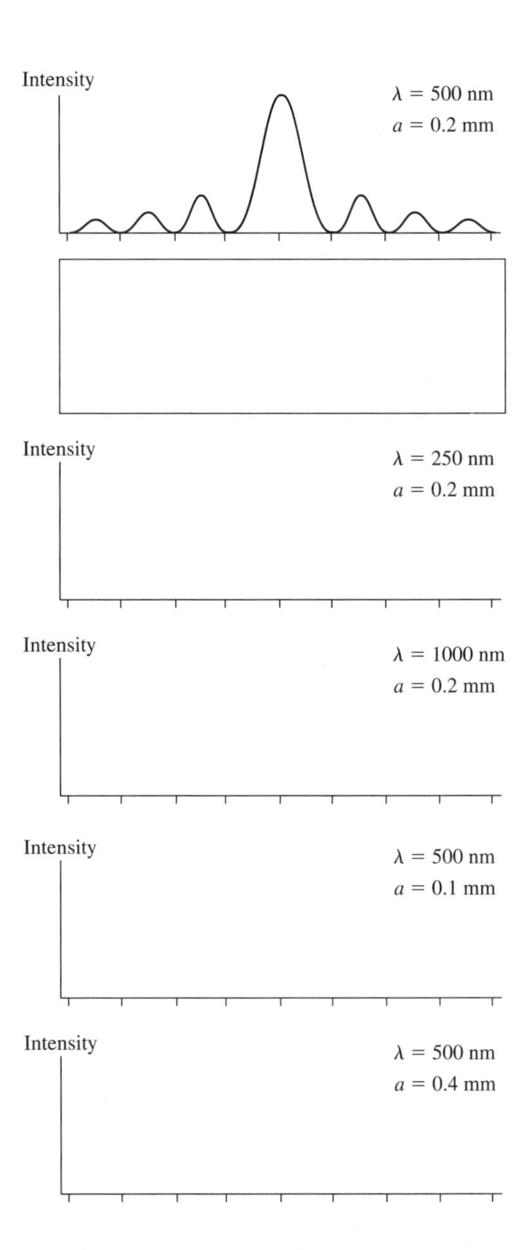

17.6 Circular-Aperture Diffraction

16. This is the light intensity on a viewing screen behind a circular aperture.

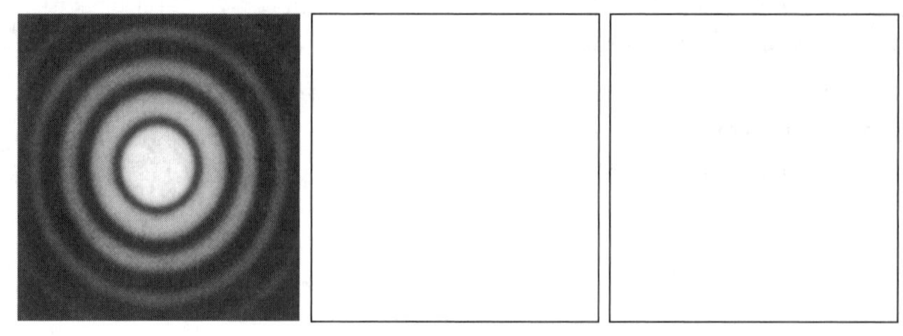

a. In the middle box, sketch how the pattern would appear if the wavelength of the light is doubled. Explain.

b. In the far right box, sketch how the pattern would appear if the diameter of the aperture is doubled. Explain.

18 Ray Optics

Note: Please use a ruler or straight edge for drawing light rays.

18.1 The Ray Model of Light

1. a. Draw four or five rays from the object that allow A to see the object.
   b. Draw four or five rays from the object that allow B to see the object.

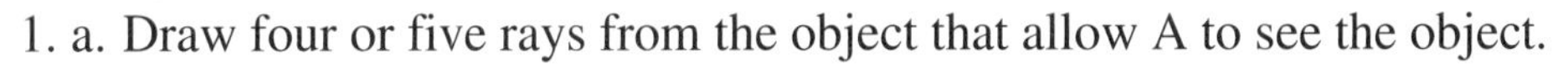
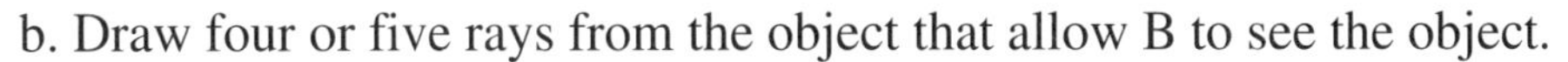

   c. Describe the situations seen by A and B if a piece of cardboard is lowered at point C.

2. a. Draw three or four rays from object 1 that allow A to see object 1.
   b. Draw three or four rays from object 2 that allow B to see object 2.
   c. What, if anything, happens to the light where the rays cross in the center of the picture?

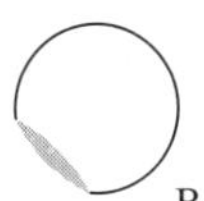

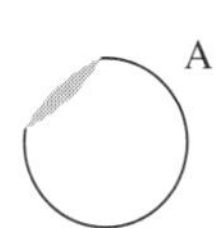

3. A point source of light illuminates a slit in an opaque barrier.
   a. On the screen, sketch the pattern of light that you expect to see. Let the white of the paper represent light areas; shade dark areas. Mark any relevant dimensions. **Note:** This slit is too wide to observe diffraction.
   b. What will happen to the pattern of light on the screen if the slit width is reduced to 0.5 cm?

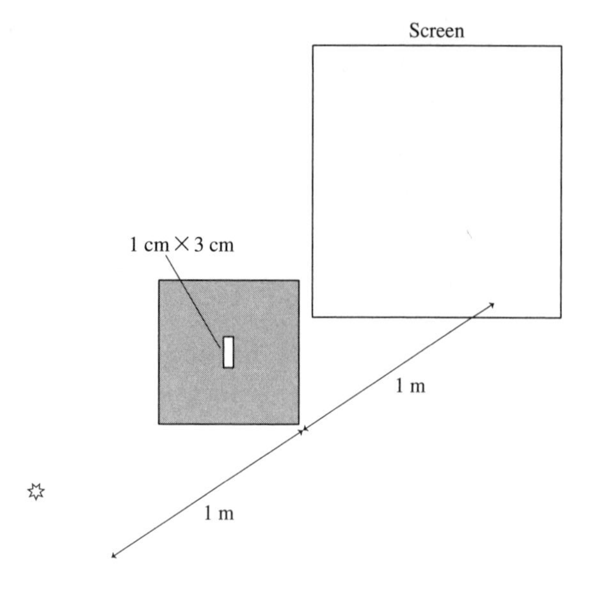

4. In each situation below, light passes through a 1-cm-diameter hole and is viewed on a screen. For each, sketch the pattern of light that you expect to see on the screen. Let the white of the paper represent light areas; shade dark areas.

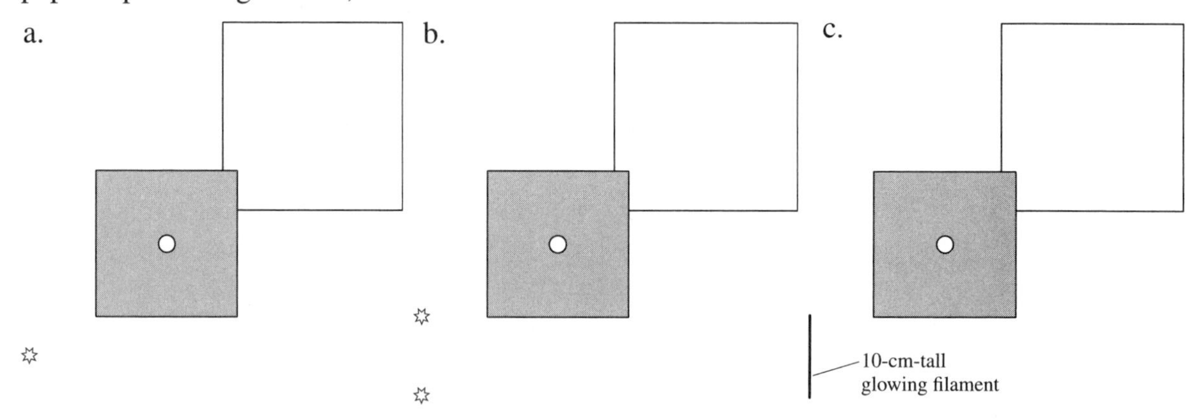

5. Light from an L-shaped bulb passes through a pinhole. On the screen, sketch the pattern of light that you expect to see.

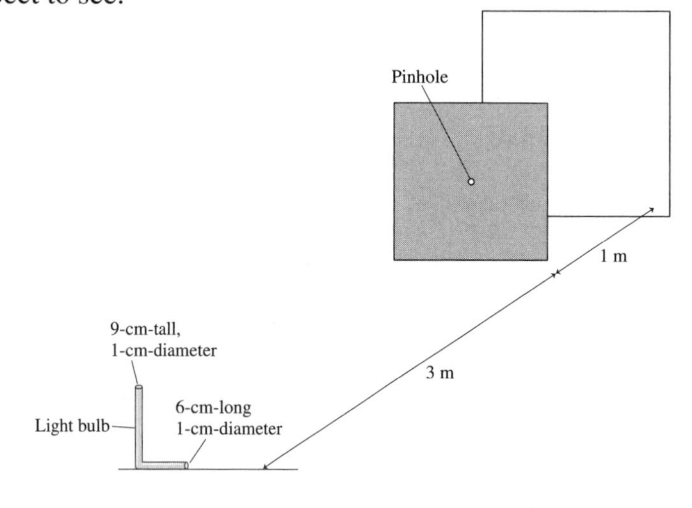

18.2 Reflection

6. a. Use a straight edge to draw five rays from the object that pass through points A to E after reflecting from the mirror. Make use of the grid to do this accurately.
   b. Extend the reflected rays behind the mirror.
   c. Show and label the image point.

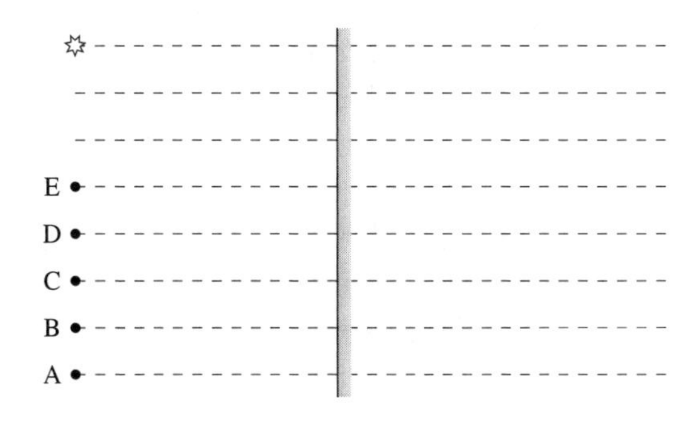

7. a. Draw *one* ray from the object that enters the eye after reflecting from the mirror.
   b. Is one ray sufficient to tell your eye/brain where the image is located?

   c. Use a different color pen or pencil to draw two more rays that enter the eye after reflecting. Then use the three rays to locate (and label) the image point.
   d. Do any of the rays that enter the eye actually pass through the image point?

8. You are looking at the image of a pencil in a mirror.
   a. What happens to the image you see if the top half of the mirror, down to the midpoint, is covered with a piece of cardboard? Explain.

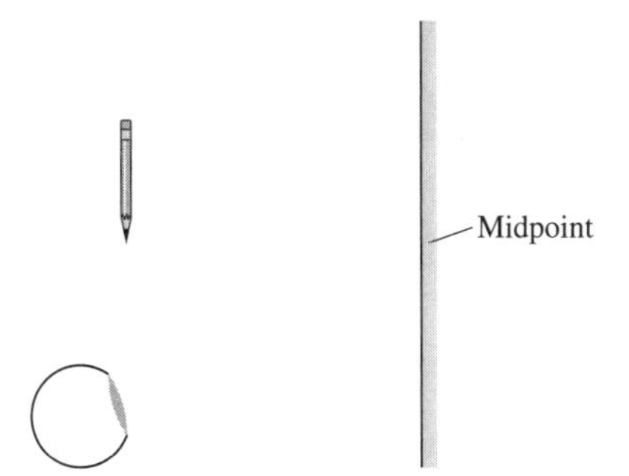

   b. What happens to the image you see if the bottom half of the mirror is covered with a piece of cardboard? Explain.

9. Two parallel mirrors face each other.
   a. Draw rays from the object that reflect from mirror 1 at the two dots. Use a straight edge and the grid to draw the reflected rays accurately.
   b. Extend the reflected rays backward to locate the object's image in mirror 1. Use a dot to indicate the image point, and label it Image 1.
   c. The rays that reflect from mirror 1 then reflect from mirror 2. Use a straight edge and the grid to accurately draw the rays reflecting from mirror 2.
   d. Extend the reflected rays backward. The point from which they appear to originate is the image, in mirror 2, of image 1. Use a dot to indicate this point, and label it Image 2.

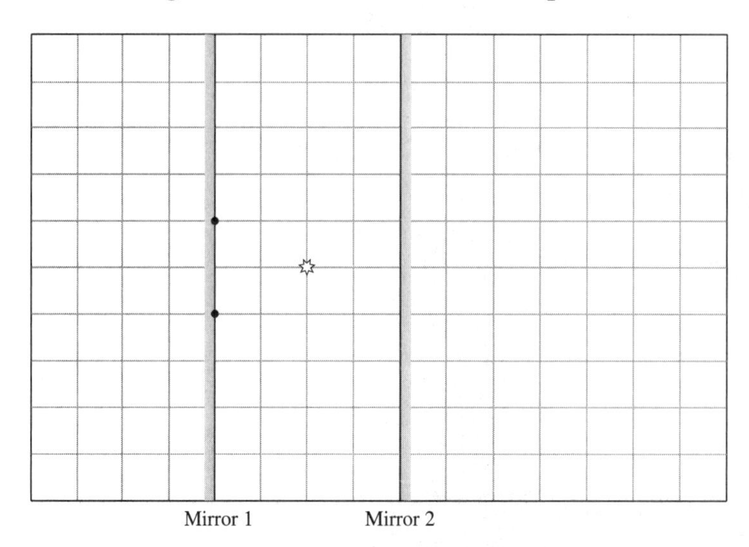

18.3 Refraction

10. Use a straight edge to complete the trajectories of these three rays through material 2 and back into material 1. Assume $n_2 < n_1$.

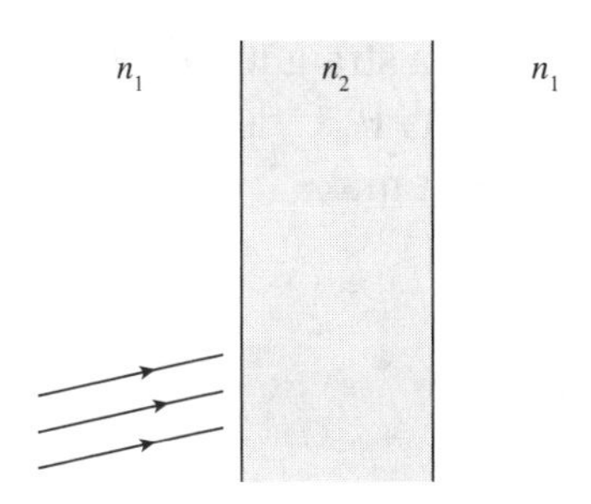

11. The figure shows six conceivable trajectories of light rays leaving an object. Which, if any, of these trajectories are impossible? For each that is possible, what are the requirements of the index of refraction n_2?

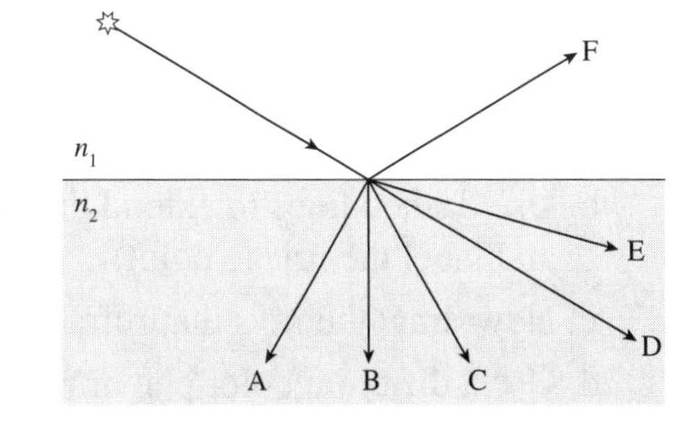

Impossible ______________

Requires $n_2 > n_1$ ______________

Requires $n_2 = n_1$ ______________

Requires $n_2 < n_1$ ______________

Possible for any n_2 ______________

12. Complete the ray trajectories through the two prisms shown below.

a.

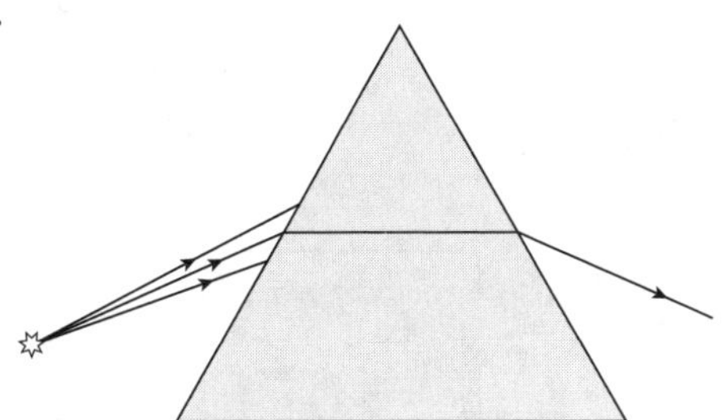

b.

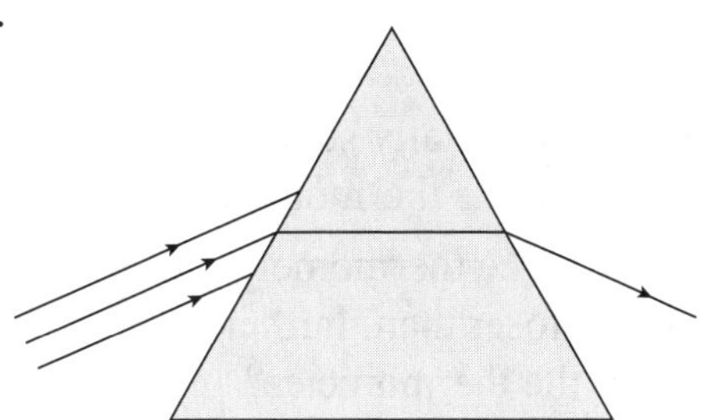

13. Draw the trajectories of seven rays that leave the object heading toward the seven dots on the boundary. Assume $n_2 < n_1$ and $\theta_c = 47°$.

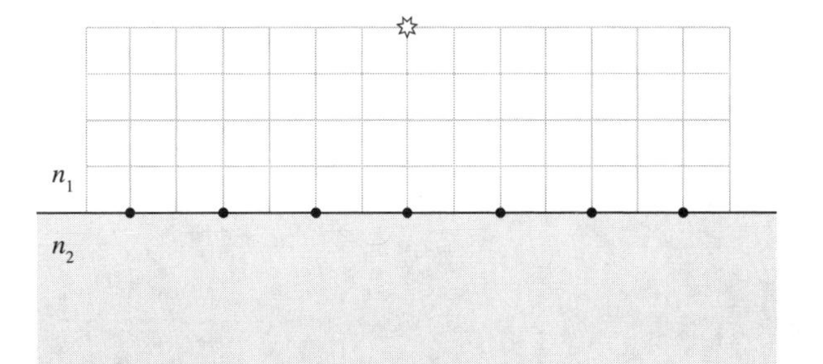

18.4 Image Formation by Refraction

14. a. Use a straight edge to draw rays that leave the object and refract after passing through points B, C, and D. Assume $n_2 > n_1$. The refraction at B and D should be the same size—don't make it too big.

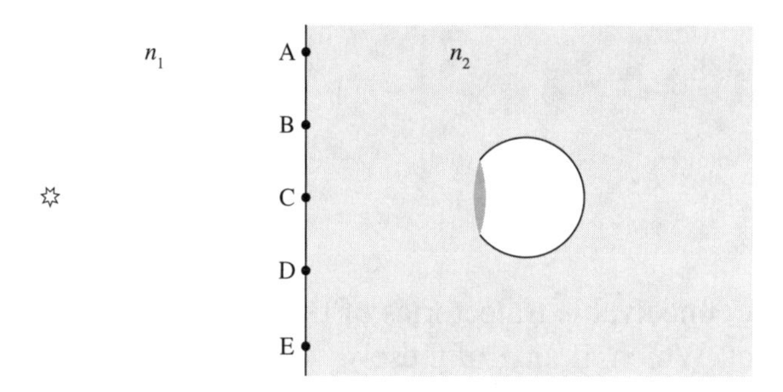

b. Use dashed lines to extend the three refracted rays backward into medium 1. Then locate and label the image point.

c. Now draw the rays that refract at A and E.

d. Use a different color pen or pencil to draw three rays from the object that enter the eye.

e. Does the distance to the object *appear* to be greater than, less than, or the same as the true distance? Explain.

15. A thermometer is partially submerged in an aquarium. The underwater part of the thermometer is not shown.

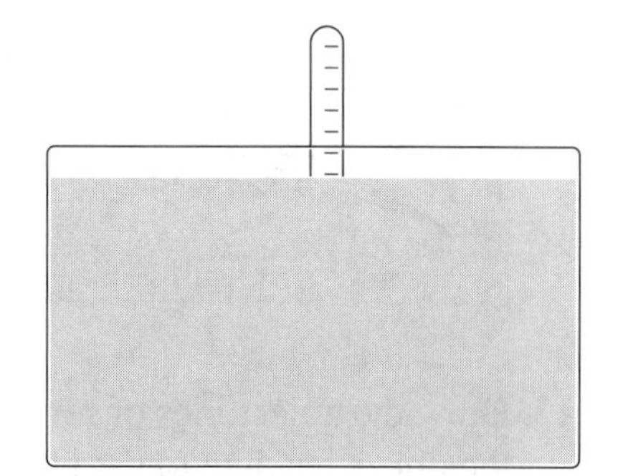

a. As you look at the thermometer, does the underwater part appear to be closer than, farther than, or the same distance as the top of the thermometer?

b. Complete the drawing by drawing the bottom of the thermometer as you think it would look.

18.5 Thin Lenses: Ray Tracing

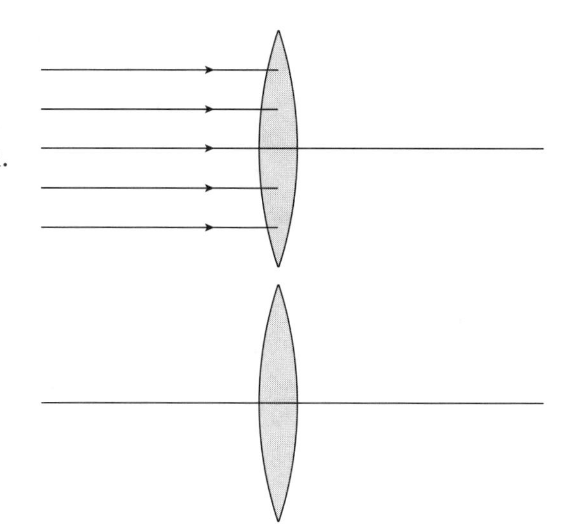

16. a. Continue these rays through the lens and out the other side.
    b. Is the point where the rays converge the same as the focal point of the lens? Or different? Explain.
    c. Place a point source of light at the place where the rays converged in part b. Draw several rays heading left, toward the lens. Continue the rays through the lens and out the other side.
    d. Do these rays converge? If so, where?

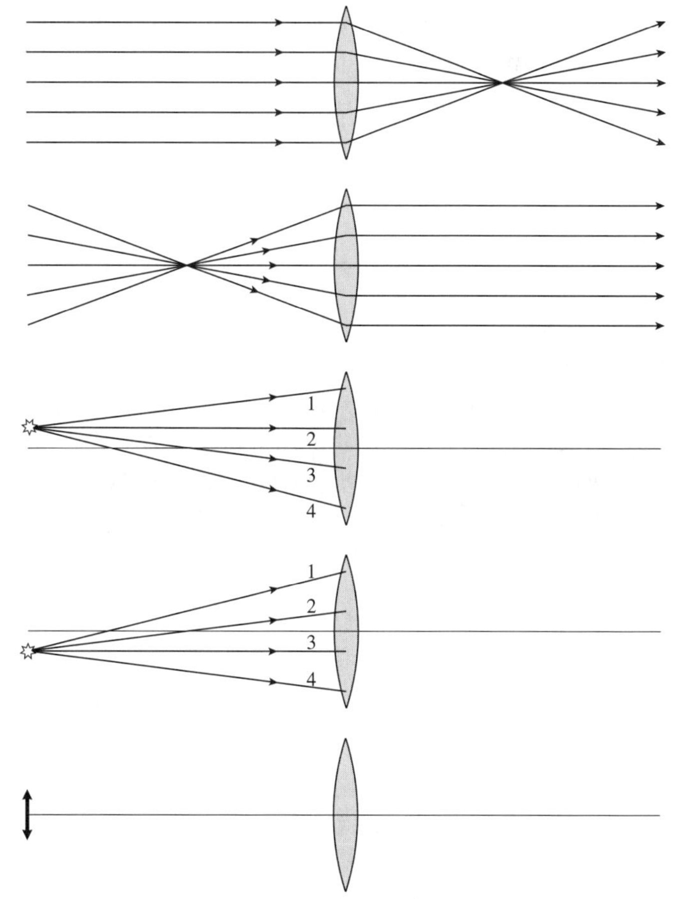

17. The top two figures show test data for a lens. The third figure shows a point source near this lens and four rays heading toward the lens.
    a. For which of these rays do you know, from the test data, its direction after passing through the lens?
    b. Draw the rays you identified in part a as they pass through the lens and out the other side.
    c. Use a different color pen or pencil to draw the trajectories of the other rays.
    d. Label the image point. What kind of image is this?
    e. The fourth figure shows a second point source. Use ray tracing to locate its image point.
    f. The fifth figure shows an extended object. Have you learned enough to locate its image? If so, draw it.

18. An object is near a lens whose focal points are marked with dots.

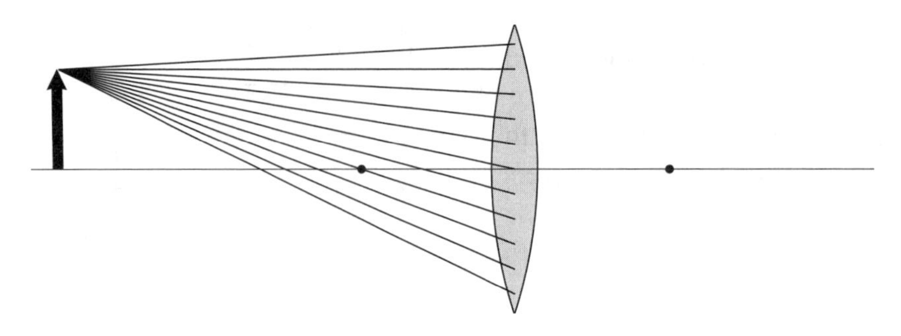

  a. Identify the three special rays and use a straight edge to continue them through the lens.
  b. Use a different color pen or pencil to draw the trajectories of the other rays.

19. An object is near a lens whose focal points are shown.
  a. Use ray tracing to locate the image of this object.
  b. Is the image upright or inverted?
  c. Is the image height larger or smaller than the object height?
  d. Is this a real or a virtual image? Explain how you can tell.

20. An object and lens are positioned to form a well-focused, inverted image on a viewing screen. Then a piece of cardboard is lowered just in front of the lens to cover the *top half* of the lens. Describe what happens to the image on the screen. What will you see when the cardboard is in place?

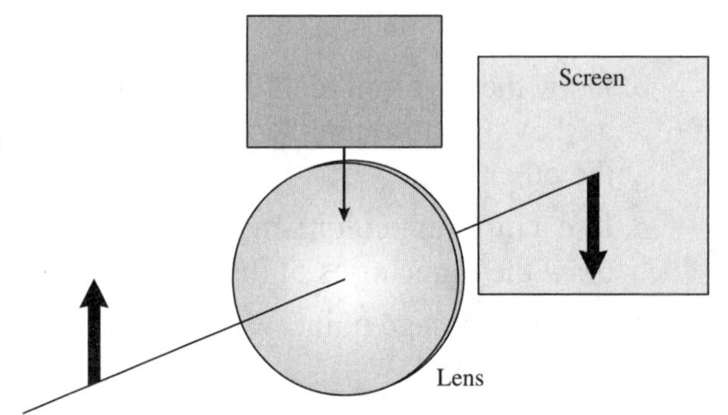

18.6 Image Formation with Spherical Mirrors

21. Two spherical mirrors are shown. The center of each sphere is marked with an open circle. For each:
    i. Use a straight edge to draw the normal to the surface at the seven dots on the boundary.
    ii. Draw the trajectories of seven rays that leave the object, strike the mirror surface at the dots, and then reflect, obeying the law of reflection.
    iii. Trace the reflected rays either forward to a point where they converge or backward to a point from which they diverge.

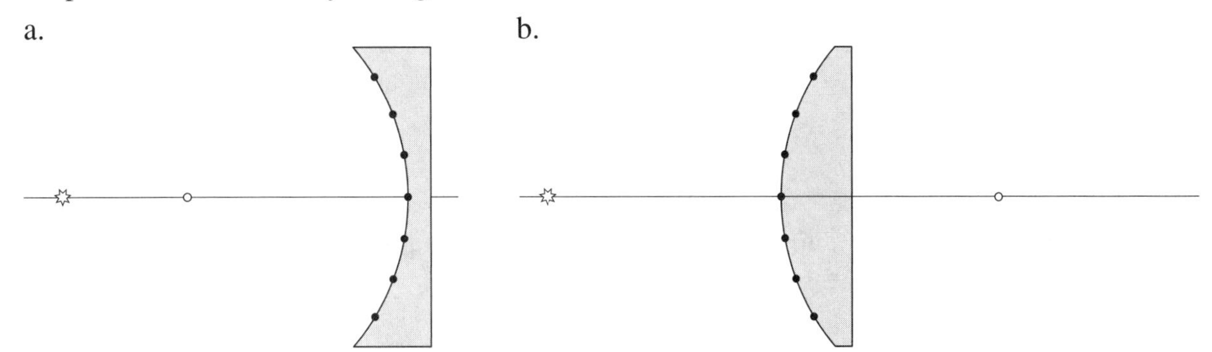

22. An object is placed near a spherical mirror whose focal point is marked.

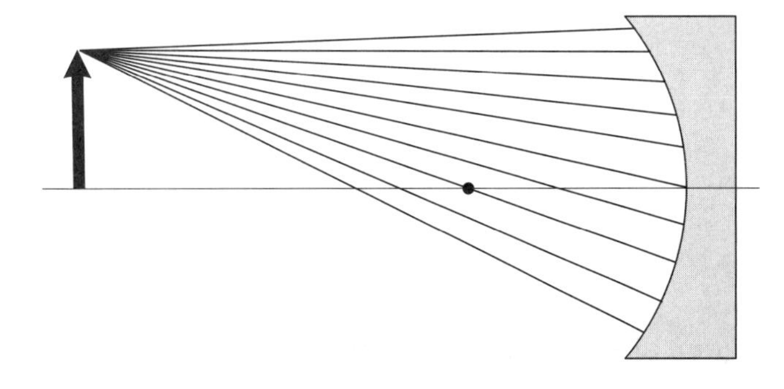

a. Identify the three special rays and show their reflections.
b. Use a different color pen or pencil to draw the trajectories of the other rays.

23. A 3.0-cm-high object is placed 10.0 cm in front of a convex diverging mirror with a focal length of −4.0 cm. Use ray tracing to determine the location of the image, the orientation of the image, and the height of the image.
    - Locate the mirror on the optical axis shown. Show its focal point with a dot.
    - Represent the object with an upright arrow at distance 10.0 cm from the mirror.
    - Draw the appropriate three "special rays" to locate the image.

18.7 The Thin-Lens Equation

24. A converging lens forms a real image. Suppose the object is moved farther from the lens. Does the image get closer to or farther from the lens? Explain.

25. A converging lens forms a virtual image. Suppose the object is moved closer to the lens. Does the image get closer to or farther from the lens? Explain.

26. The figure is a ray diagram of image formation by a mirror. In this situation,

    Is s positive, negative, or zero? ______________

    Is s' positive, negative, or zero? ______________

    Is f positive, negative, or zero? ______________

    Is m positive, negative, or zero? ______________

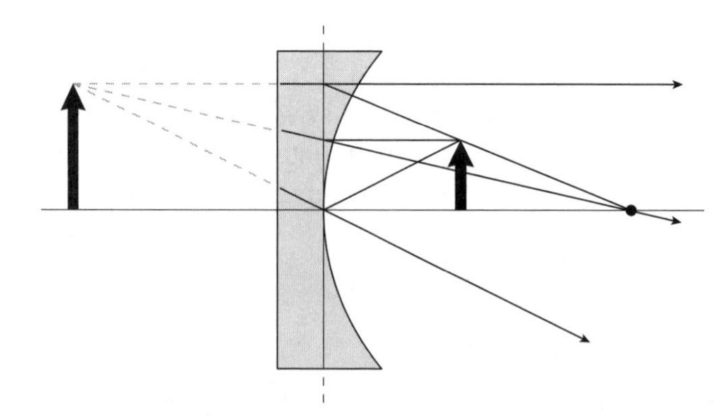

19 Optical Instruments

19.1 The Camera

19.2 The Human Eye

1. A photographer focuses his camera on an object. Suppose the object moves closer to the camera. To refocus, should the camera lens move closer to or farther from the detector? Explain.

2. Two lost students wish to start a fire to keep warm while they wait to be rescued. One student is hyperopic, the other myopic. Which, if either, could use his glasses to focus the sun's rays to an intense bright point of light? Explain.

3. Suppose you wanted special glasses designed to let you see underwater, without a face mask. Should the glasses use a converging or diverging lens? Explain.

4. Equip each eyeball below with an appropriate eyeglass lens that will produce a well-focused image on the retina. To do so, first draw the lens in front of the eye, then redraw the two rays from the point at which they enter the lens until they form an image.

a.

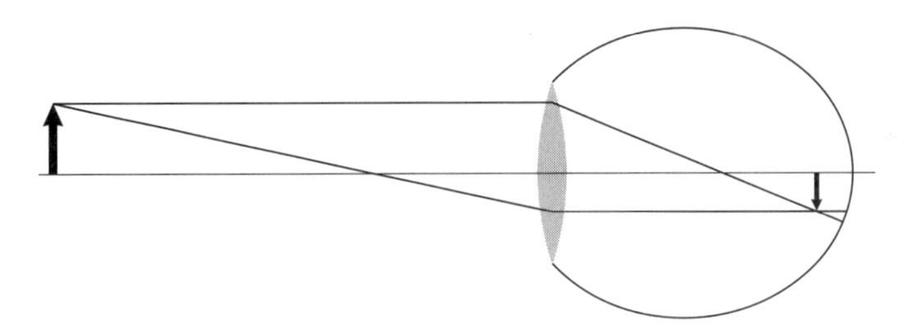

b. Is this eye hyperopic or myopic? ______________ Explain.

c.

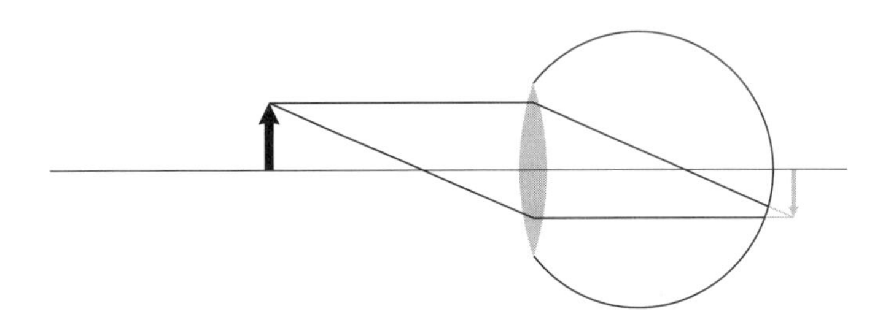

d. Is this eye hyperopic or myopic? ______________ Explain.

19.3 The Magnifier

5. An eye views objects A and B.

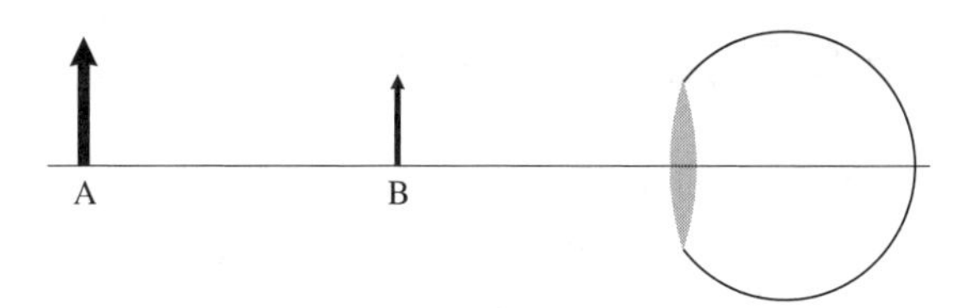

   a. Which object has the larger size? Explain.

   b. Which object has the larger angular size? Explain.

6. The angular magnification of a lens is not sufficient. To double the angular magnification, do you want a lens with twice the focal length or half the focal length? Explain.

7. On the left, an eye observes an object at the eye's near point of 25 cm. This is the closest the object can be and still be seen clearly. On the right, the eye views the same object through a magnifying lens. The object's physical distance from the eye is now much less than 25 cm.

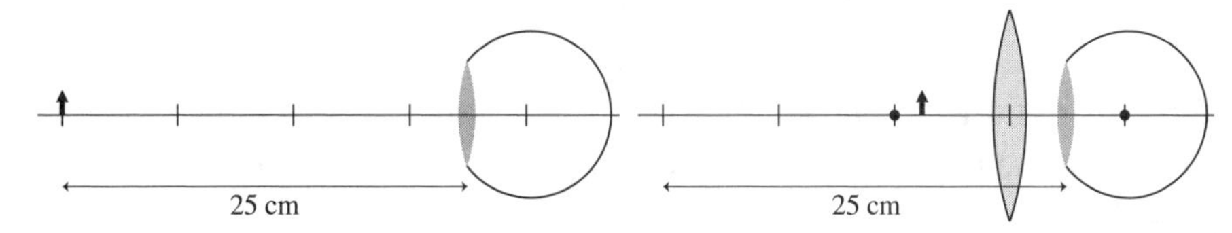

   a. Using a straight edge, draw a line on the left figure to indicate the angular size of the object when viewed by the unaided eye at the eye's near point.
   b. Use ray tracing with a straight edge to show that the image in the right figure is a virtual image ≈ 25 cm from the eye.
   c. Draw a line (or label one of your ray-tracing lines) to indicate the angular size of the image seen through the lens.
   d. Using a ruler to make measurements, determine the angular magnification of the lens.

   e. It's sometimes said that a magnifying glass makes an object "appear closer." Is that what is happening here? Explain.

19.4 The Microscope

19.5 The Telescope

8. a. Complete the ray diagram by drawing two special rays that start from the tip of the objective's image and pass through the eyepiece. Show that the two rays are parallel on the right side of the eyepiece. (Because these rays are parallel, it is not possible to draw the final virtual image on your diagram.)

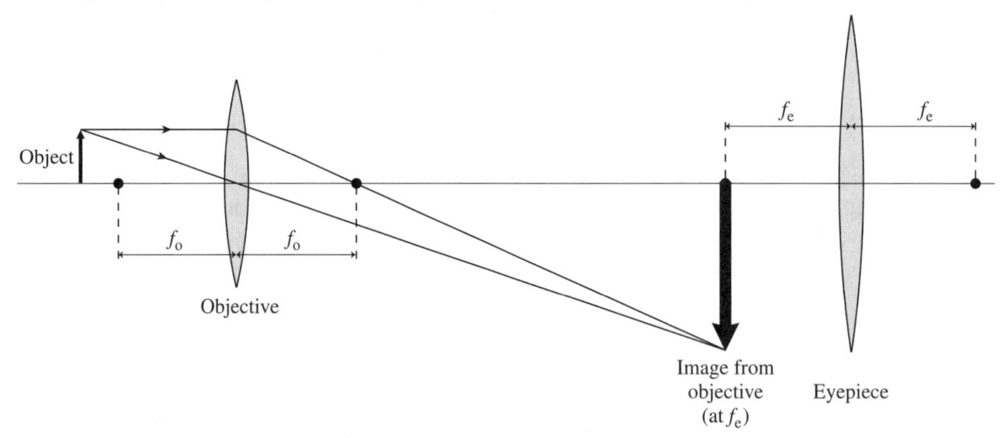

b. On the diagram above, indicate the angle subtended by the final image. This is the image's angular size. Measure this angle with a protractor.

θ final image = ____________

c. Draw a line from far right end of the axis, where your eye would be placed, to the tip of the object. Measure the angle of this line with a protractor. This is the object's angular size.

θ object = ____________

d. What is the angular magnification of the two-lens system?

9. Rank in order, from largest to smallest, the magnifications M_1 to M_4 of these telescopes.

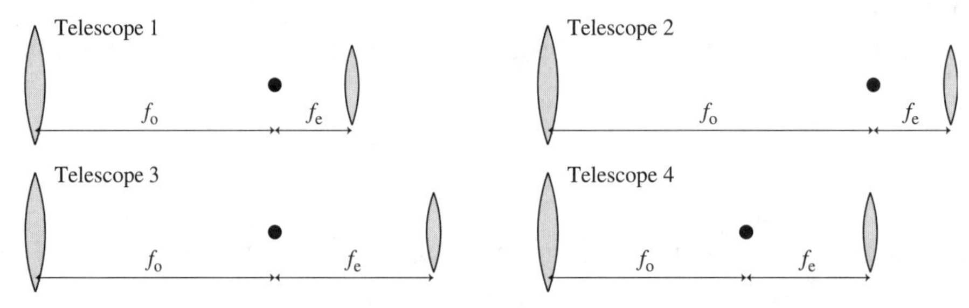

Order:

Explanation:

19.6 Color and Dispersion

10. A beam of white light from a flashlight passes through a red piece of plastic.
    a. What is the color of the light that emerges from the plastic?
    b. Is the emerging light as intense as, more intense than, or less intense than the white light? Explain.

    c. The light then passes through a blue piece of plastic. Describe the color and intensity of the light that emerges.

11. Suppose you looked at the sky on a clear day through pieces of red and blue plastic oriented as shown. Describe the color and brightness of the light coming through sections 1, 2, and 3.

Red
3
2
1
Blue

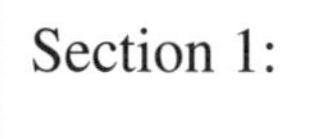

Section 1:

Section 2:

Section 3:

12. Sketch a plausible absorption spectrum for a patch of bright red paint.

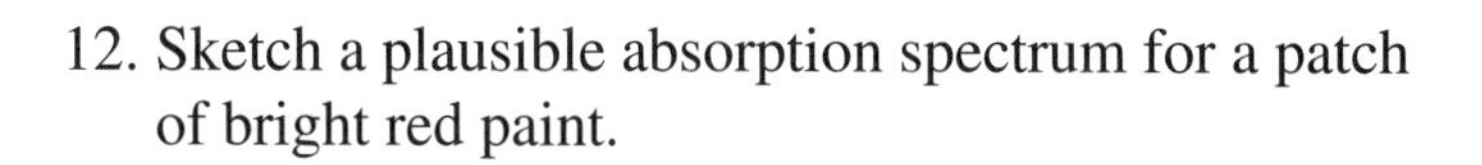

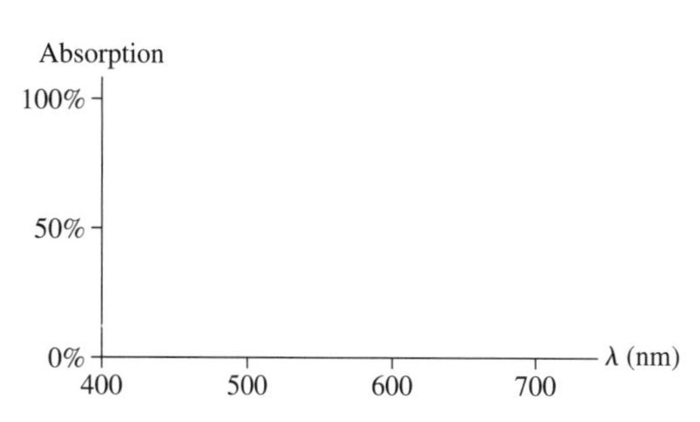

19.7 Resolution of Optical Instruments

13. A diffraction-limited lens can focus light to a 10-μm-diameter spot on a screen. Do the following actions make the spot diameter larger, smaller, or leave it unchanged?

    a. Decreasing the wavelength of light: ____________

    b. Decreasing the lens diameter: ____________

    c. Decreasing the focal length: ____________

    d. Decreasing the lens-to-screen distance: ____________

14. An astronomer is trying to observe two distant stars. The stars are marginally resolved when she looks at them through a filter that passes green light near 550 nm. Which of the following actions would improve the resolution? Assume that the resolution is not limited by the atmosphere.

    a. Changing the filter to a different wavelength? If so, should she use a shorter or a longer wavelength?

    b. Using a telescope with an objective lens of the same diameter but different focal length? If so, should she select a shorter or a longer focal length?

    c. Using a telescope with an objective lens of the same focal length but a different diameter? If so, should she select a larger or a smaller diameter?

    d. Using an eyepiece with a different magnification? If so, should she select an eyepiece with more or less magnification?

20 Electric Fields and Forces

20.1 Charges and Forces

1. Two lightweight balls hang straight down when both are neutral. They are close enough together to interact, but not close enough to touch. Draw pictures showing how the balls hang if:

A B

   a. Both are touched with a plastic rod that was rubbed with wool.

   b. The two charged balls of part a are moved farther apart.

   c. Ball A is touched by a plastic rod that was rubbed with wool and ball B is touched by a glass rod that was rubbed with silk.

   d. Both are charged by a plastic rod, but ball A is charged more than ball B.

   e. Ball A is charged by a plastic rod. Ball B is neutral.

2. After combing your hair briskly, the comb will pick up small pieces of paper.
   a. Is the comb charged? Explain.

   b. How can you be sure that it isn't the paper that is charged? Propose an experiment to test this.

   c. Is your hair charged after being combed? What evidence do you have for your answer?

   d. What kind of charge—positive or negative—is the comb likely to have? Why?

   e. How could you test your answer to part d?

3. A negatively charged electroscope has separated leaves.
   a. Suppose you bring a negatively charged rod close to the top of the electroscope, but not touching. How will the leaves respond? Use both charge diagrams and words to explain.
   b. How will the leaves respond if you bring a positively charged rod close to the top of the electroscope, but not touching? Use both charge diagrams and words to explain.

a.

b.

4. Four lightweight balls A, B, C, and D are suspended by threads. Ball A has been touched by a plastic rod that was rubbed with wool. When the balls are brought close together, without touching, the following observations are made:

- Balls B, C, and D are attracted to ball A.
- Balls B and D have no effect on each other.
- Ball B is attracted to ball C.

What are the charge states (positive, negative, or neutral) of balls A, B, C, and D? Explain.

5. a. Metal sphere A is initially neutral. A positively charged rod is brought near, but not touching. Is A now positive, negative, or neutral? Explain.

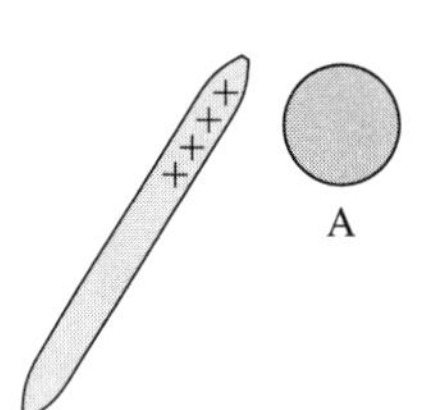

b. Metal spheres A and B are initially neutral and are touching. A positively charged rod is brought near A, but not touching. Is A now positive, negative, or neutral? Explain.

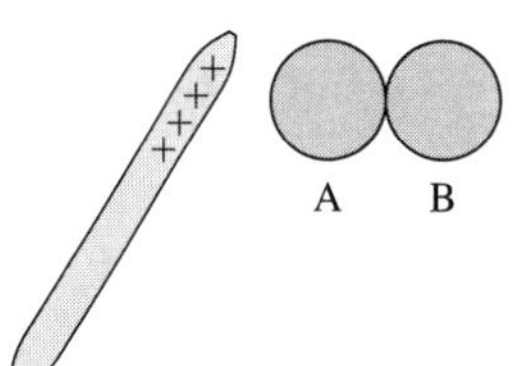

c. Metal sphere A is initially neutral. It is connected by a metal wire to the ground. A positively charged rod is brought near, but not touching. Is A now positive, negative, or neutral? Explain.

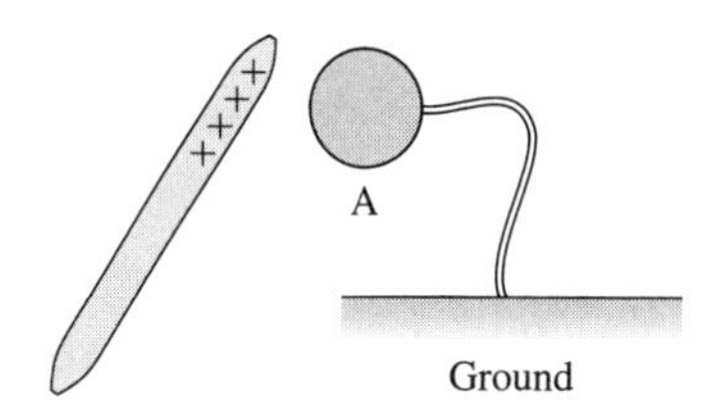

20.2 Charges, Atoms, and Molecules

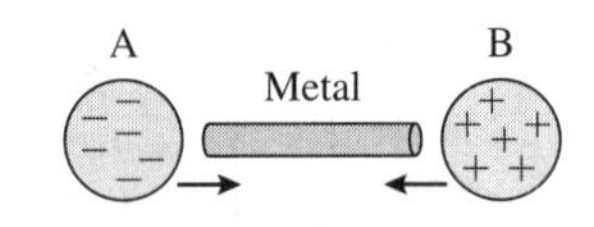

6. Two oppositely charged metal spheres have equal quantities of charge. They are brought into contact with a neutral metal rod.

   a. What is the final charge state of each sphere and of the rod?

   b. Use both charge diagrams and words to explain how this charge state is reached.

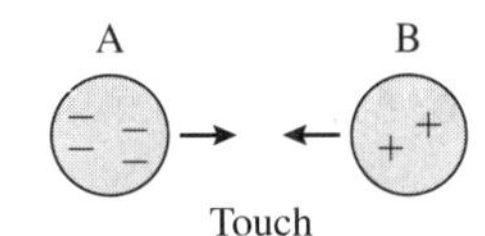

7. Metal sphere A has 4 units of negative charge and metal sphere B has 2 units of positive charge. The two spheres are brought into contact. What is the final charge state of each sphere? Explain.

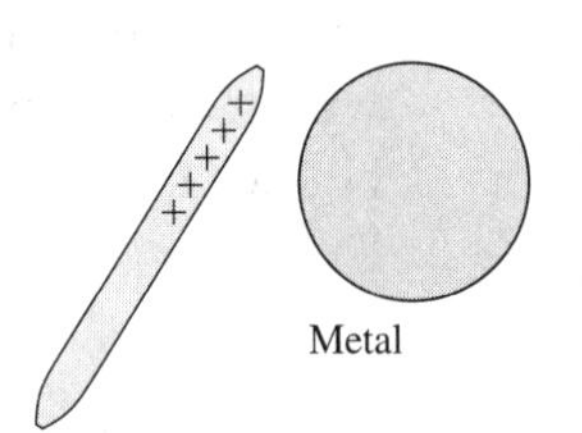

8. A positively charged rod is held near, but not touching, a neutral metal sphere.

   a. Add plus and minus signs to the figure to show the charge distribution on the sphere.

   b. Does the sphere experience a net force? If so, in which direction? Explain.

20.3 Coulomb's Law

9. For each pair of charges, draw a force vector *on each charge* to show the electric force acting on that charge. The length of each vector should be proportional to the magnitude of the force. Each + and – symbol represents the same quantity of charge.

a. (−) (−)

b. (++) (−)

c. (+++) (−)

d. (+++) (++)

10. For each group of charges, use a **black** pen or pencil to draw each force acting on the gray positive charge. Then use a **red** pen or pencil to show the net force on the gray charge. Label $\vec{F}_{net}$.

a. (+) (+) (+)

b. (−) (+) (+)

c. (++) (+) (−)

11. Can you assign charges (positive or negative) so that these forces are correct? If so, show the charges on the figure. (There may be more than one correct response.) If not, why not?

a.

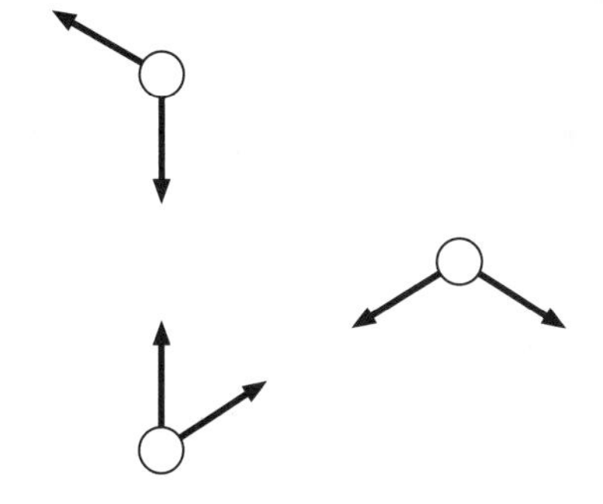

b.

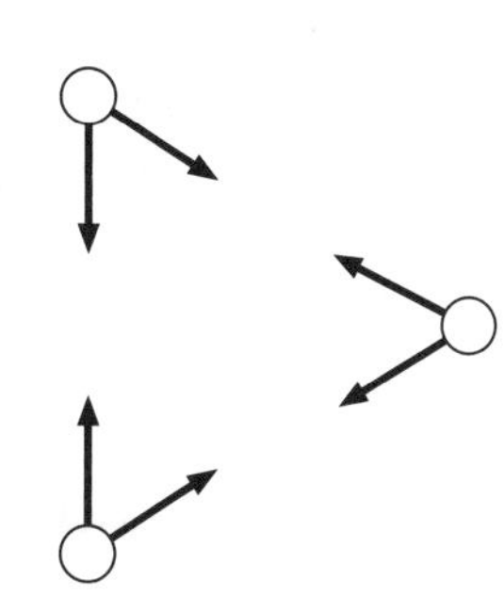

c.

d.

12. Draw a + on the figure below to show the position or positions where a proton would experience no net force.

13. Draw a − on the figure below to show the position or positions where an electron would experience no net force.

14. The gray positive charge experiences a net force due to two other charges: the +1 charge that is seen and a +4 charge that is not seen. Add the +4 charge to the figure at the correct position.

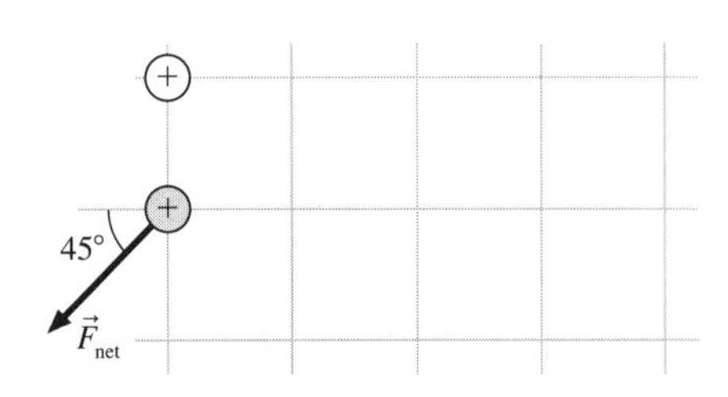

15. Positive charges $4q$ and q are distance L apart. Let them be on the x-axis with $4q$ at the origin.

PSA 20.1

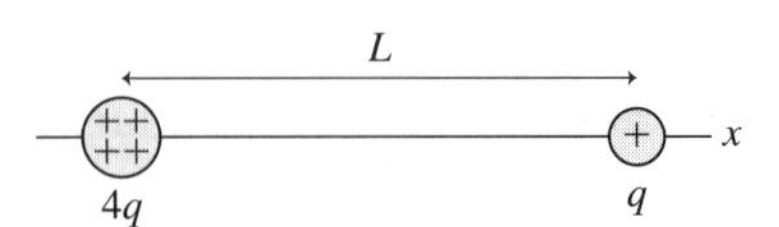

a. Suppose a proton were placed on the x-axis to the *right* of q. Is it *possible* that the net electric force on the proton is zero? Explain.

b. On the figure, draw a proton at an arbitrary point on the x-axis between $4q$ and q. Label its distance from $4q$ as r. Draw two force vectors and label them $\vec{F}_{4q}$ and $\vec{F}_q$ to show the two forces on this proton. Is it *possible* that, for the proper choice of r, the net electric force on the proton is zero? Explain.

c. Write expressions for the magnitudes of forces $\vec{F}_{4q}$ and $\vec{F}_q$ Your expressions should be in terms of K, q, L, and r.

$F_{4q} =$ ______ $F_q =$ ______

d. Find the specific value of r—as a fraction of L—at which the net force is zero.

20.4 The Concept of the Electric Field

16. At points 1 to 4, draw an electric field vector with the proper direction and with a length proportional to the electric field strength at that point.

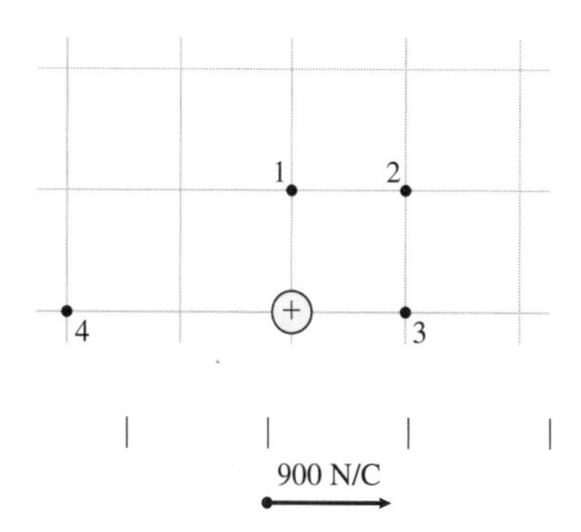

17. a. The electric field of a point charge is shown at *one* point in space.
    Can you tell if the charge is + or −? If not, why not?

900 N/C

b. Here the electric field of a point charge is shown at two positions in space.
   Now can you tell if the charge is + or −? Explain.

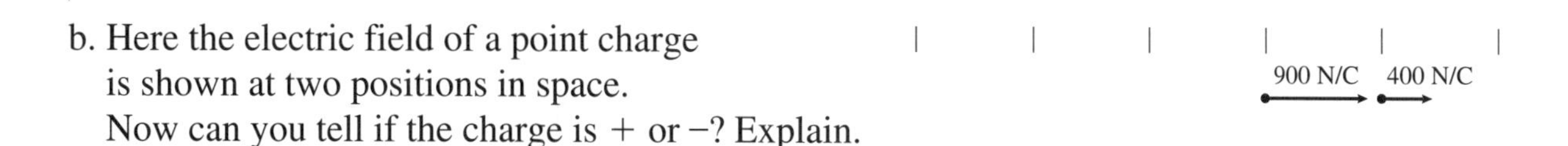

c. Can you determine the location of the charge? If so, draw it on the figure. If not, why not?

18. The electric field strength at a point in space near a point charge is 1000 N/C.
    a. What will be the field strength if the quantity of charge is halved? Explain.

    b. What will be the field strength if the distance to the point charge is halved? The quantity of charge is the original amount, not the value of part a. Explain.

19. Is the electric field strength at point A larger than, smaller than, or the same as the electric field strength at point B? Explain.

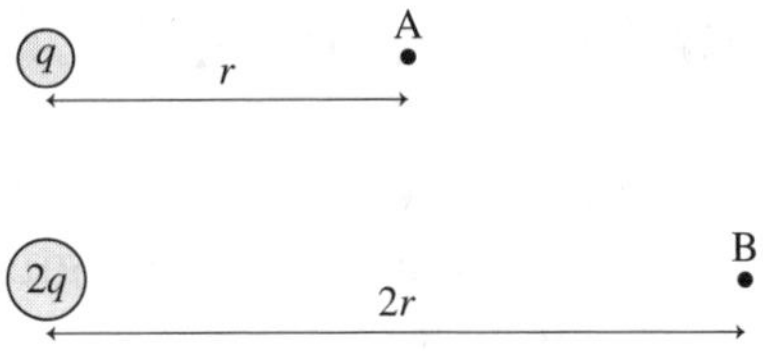

20. Is there an electric field at the position of the dot? If so, draw the electric field vector on the figure. If not, what would you need to do to create an electric field at this point?

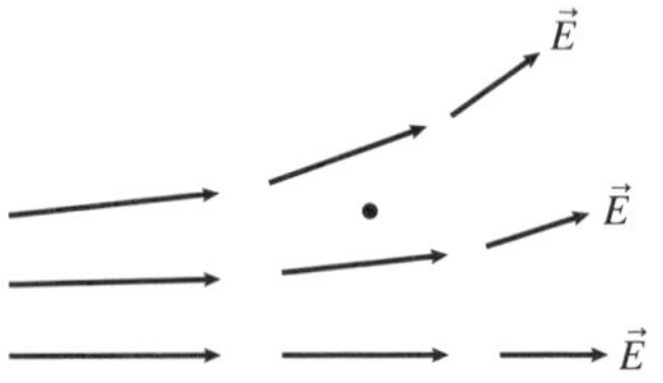

20.5 The Electric Field of Multiple Charges

21. At each of the dots, use a **black** pen or pencil to draw and label the electric fields $\vec{E}_1$ and $\vec{E}_2$ due to the two point charges. Make sure that the *relative* lengths of your vectors indicate the strength of each electric field. Then use a **red** pen or pencil to draw and label the net electric field $\vec{E}_{net}$.

a.

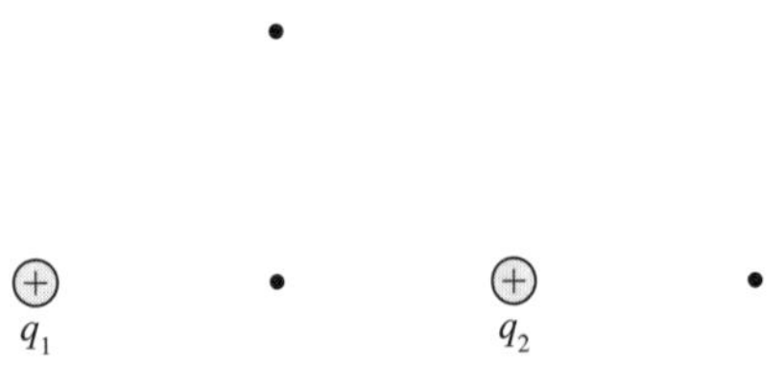

b.

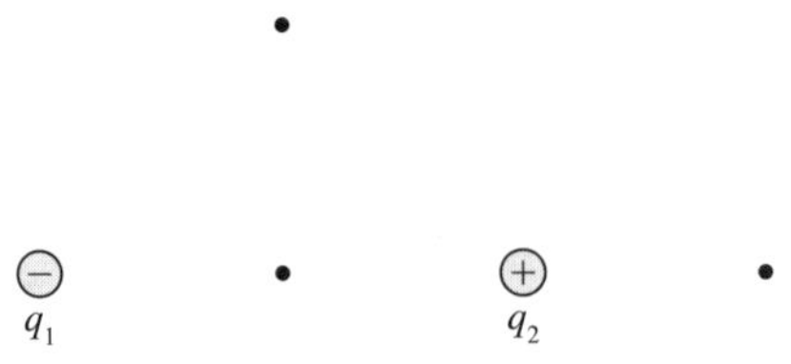

22. For each of the figures, use dots to mark any point or points (other than infinity) where $\vec{E} = \vec{0}$.

a.

b.

23. The figure shows the electric field lines in a region of space. Draw the electric field vectors at the three dots.

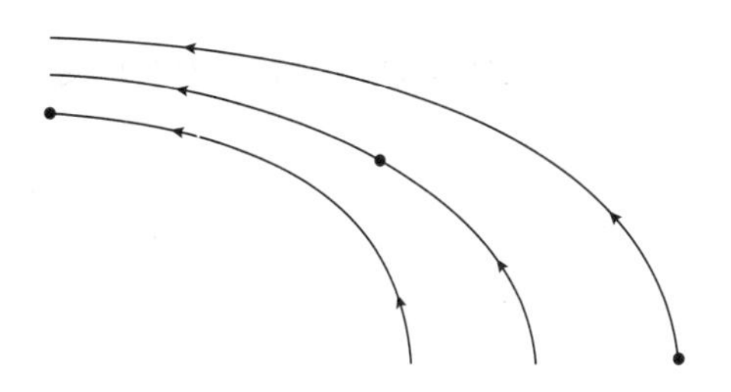

24. Rank in order, from largest to smallest, the electric field strengths E_1 to E_5 at each of these points.

Order:

Explanation:

25. A parallel-plate capacitor is constructed of two square plates, size $L \times L$, separated by distance d. The plates are given charge $\pm Q$. What is the ratio E_f/E_i of the final electric field strength E_f to the initial electric field strength E_i if:

a. Q is doubled?

b. L is doubled?

c. d is doubled?

20.6 Conductors in Electric Fields

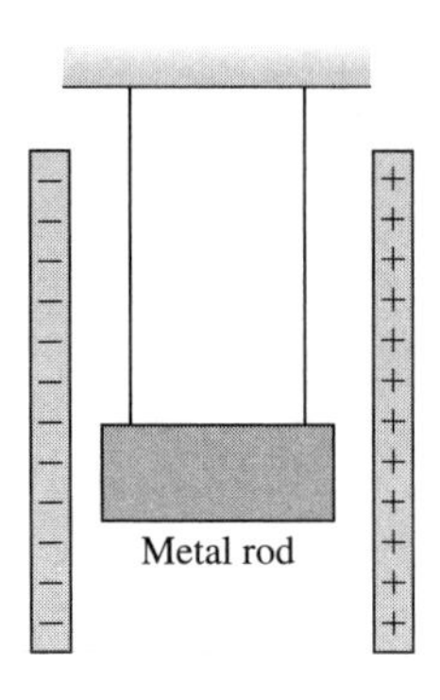

26. A neutral metal rod is suspended in the center of a parallel-plate capacitor. Then the capacitor is charged as shown.
    a. Is the rod now positive, negative, or neutral? Explain.

    b. Is the rod polarized? If so, draw plus and minus signs on the figure to show the charge distribution. If not, why not?

    c. Does the rod swing toward one of the plates, or does it remain in the center? If it swings, which way? Explain.

20.7 Forces and Torques in Electric Fields

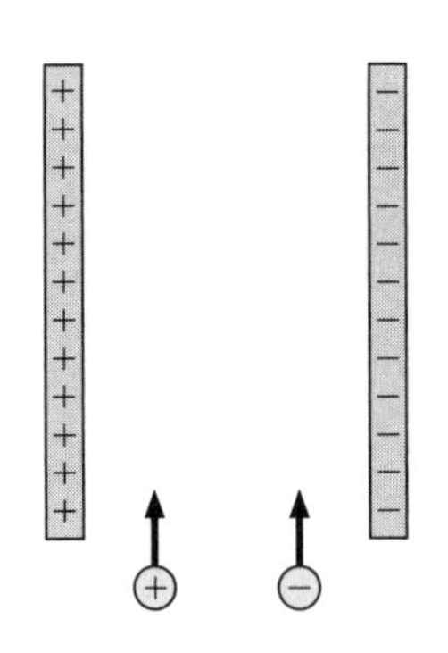

27. Positively and negatively charged particles, with equal masses and equal quantities of charge, are shot into a capacitor in the directions shown.
    a. Use solid lines to draw their trajectories on the figure if their initial velocities are fast.
    b. Use dotted lines to draw their trajectories on the figure if their initial velocities are slow.

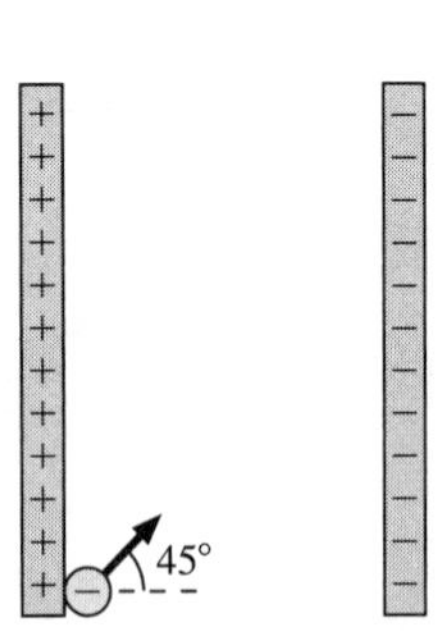

28. An electron is launched from the positive plate at a 45° angle. It does not have sufficient speed to make it to the negative plate. Draw its trajectory on the figure.

29. Three charges are placed at the corners of a triangle. The ++charge has twice the quantity of charge of the two – charges; the net charge is zero.

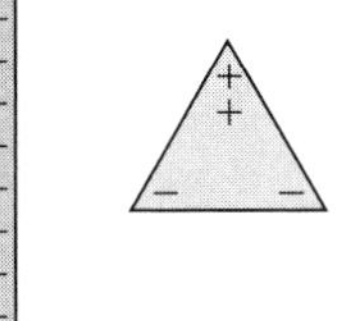

   a. Draw the force vectors on each of the charges.
   b. Is the triangle in equilibrium? ______________ If not, draw the equilibrium orientation directly beneath the triangle that is shown.
   c. Once in the equilibrium orientation, will the triangle move to the right, move to the left, rotate steadily, or be at rest? Explain.

You Write the Problem!

Exercises 30–32: You are given the equation that is used to solve a problem. For each of these:

a. Write a *realistic* physics problem for which this is the correct equation. Look at worked examples and end-of-chapter problems in the textbook to see what realistic physics problems are like.
b. Finish the solution of the problem.

30. $$\frac{(9.0 \times 10^9\ \text{N}\cdot\text{m}^2/\text{C}^2) \times q^2}{(0.015\ \text{m})^2} = 0.020\ \text{N}$$

31. $$\frac{(9.0 \times 10^9\ \text{N}\cdot\text{m}^2/\text{C}^2) \times N \times (1.6 \times 10^{-19}\ \text{C})}{(1.0 \times 10^{-6}\ \text{m})^2} = 1.5 \times 10^6\ \text{N/C}$$

32. $$1.5 \times 10^6\ \text{N/C} = \frac{Q}{(8.85 \times 10^{-12}\ \text{C}^2/\text{N}\cdot\text{m}^2)\,\pi\,(0.0125\ \text{m})^2}$$

21 Electric Potential

21.1 Electric Potential Energy and Electric Potential

1. A force does 2 μJ of work to push charged particle A toward a set of fixed source charges. Charged particle B has twice the charge of A. How much work must the force do to push B through the same displacement? Explain.

2. A 1 nC charged particle is pushed toward a set of fixed source charges, as shown. In the process, the particle gains 1 μJ of electric potential energy.

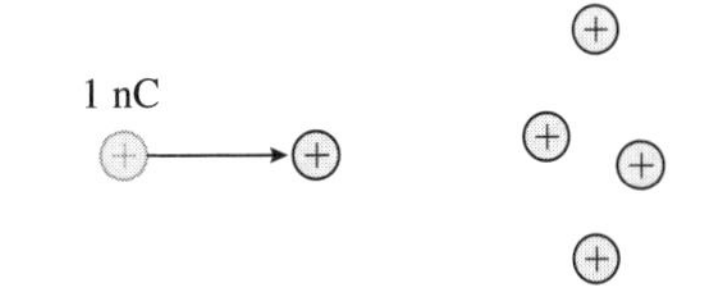

   a. How much work was done to push the particle through this displacement? Explain.

   b. Through what potential difference did the particle move?

3. Charged particle A is placed at a point in space where the electric potential is V. Its electric potential energy at that point is U_A. Particle A is removed and replaced by charged particle B, whose potential energy at the same point is U_B. If the charge of B is three times the charge of A, what is the ratio U_B/U_A? Explain.

4. Which point, A or B, has the higher electric potential? Why?

21.2 Sources of Electric Potential

21.3 Electric Potential and Conservation of Energy

5. A positive charge q is fired through a small hole in the positive plate of a capacitor. Does q speed up or slow down inside the capacitor? Answer this question twice:

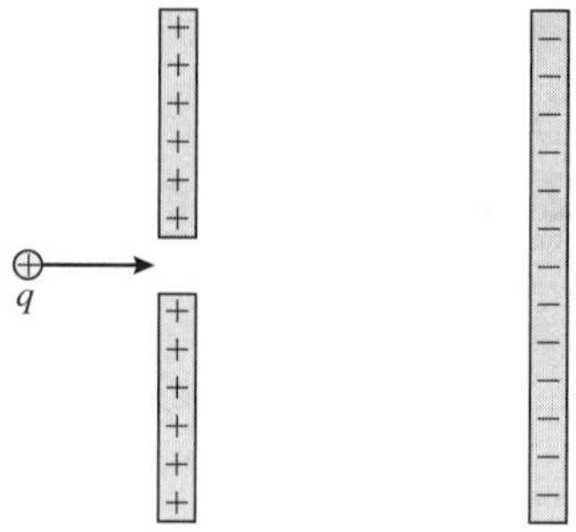

   a. First using the concept of force.

   b. Second using the concept of energy.

6. Charge q is fired toward a stationary positive point charge.

   a. If q is a positive charge, does it speed up or slow down as it approaches the stationary charge? Answer this question twice:

      i. Using the concept of force.

      ii. Using the concept of energy.

   b. Repeat part a for q as a negative charge.

21.4 Calculating the Electric Potential

7. Rank in order, from largest to smallest, the electric potentials V_1 to V_5 at points 1 to 5.

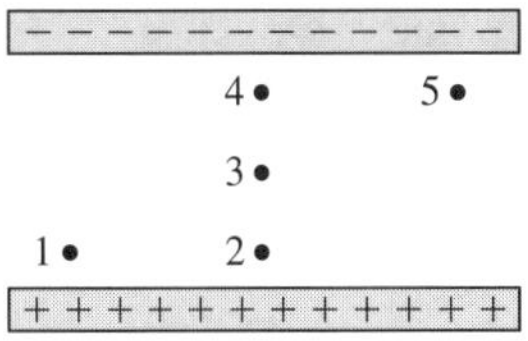

Order:

Explanation:

8. The figure shows two points inside a capacitor. Let $V = 0$ V at the negative plate.

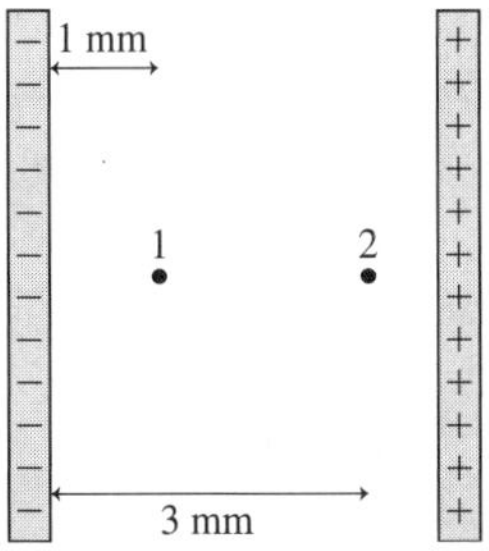

a. What is the ratio V_2/V_1 of the electric potentials at these two points? Explain.

b. What is the ratio E_2/E_1 of the electric field strengths at these two points? Explain.

9. A capacitor with plates separated by distance d is charged to a potential difference ΔV_C. All wires and batteries are disconnected, and then the two plates are pulled apart (with insulated handles) to a new separation of distance $2d$.

a. Does the capacitor charge Q change as the separation increases? If so, by what factor? If not, why not?

b. Does the electric field strength E change as the separation increases? If so, by what factor? If not, why not?

c. Does the potential difference ΔV_C change as the separation increases? If so, by what factor? If not, why not?

10. Each figure shows a contour map on the left and a set of graph axes on the right. Draw a graph of V versus x. Your graph should be a straight line or a smooth curve.

a.

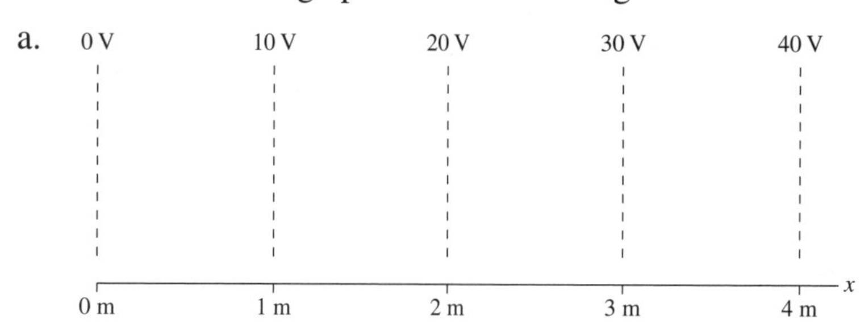

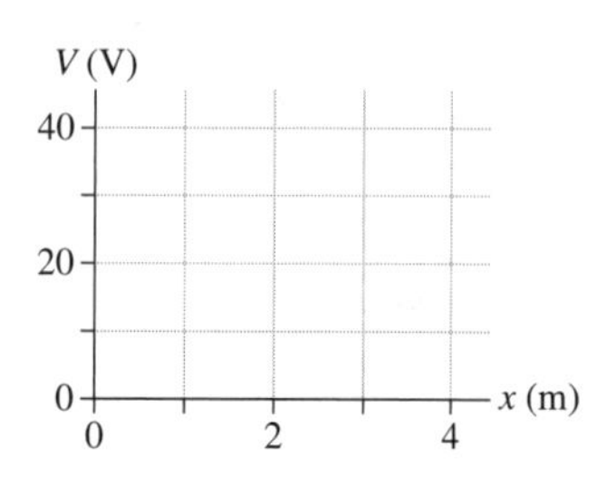

b.

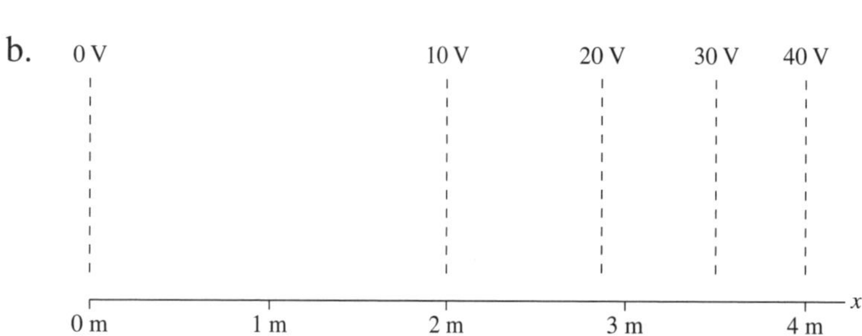

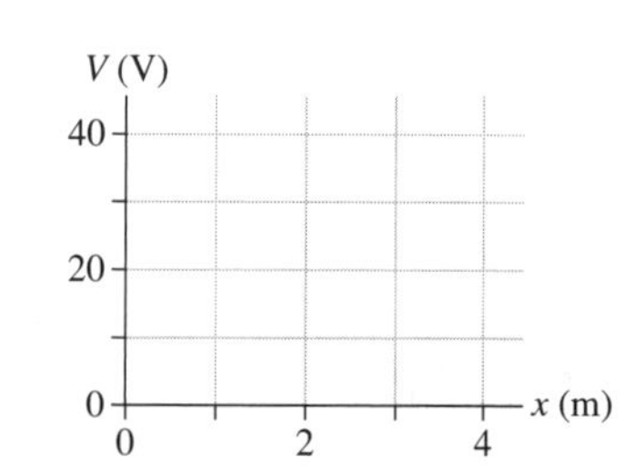

c.

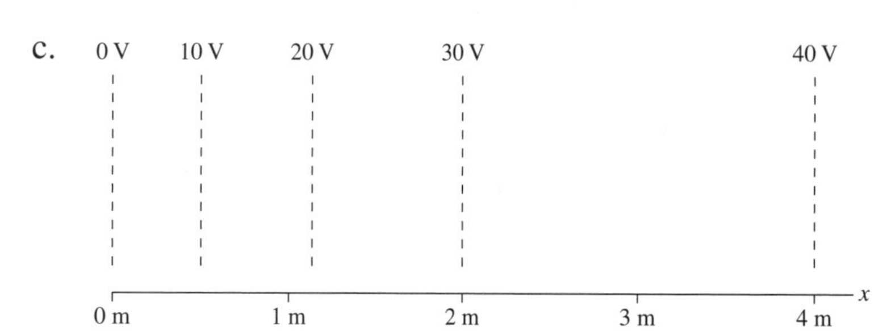

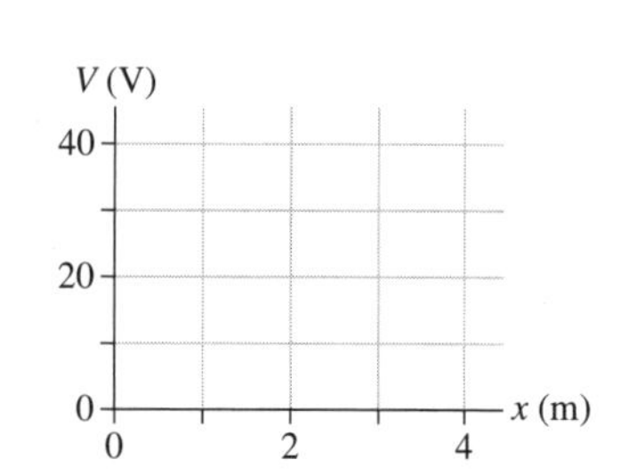

11. Each figure shows a V-versus-x graph on the left and an x-axis on the right. Assume that the potential varies with x but not with y. Draw a contour map of the electric potential. There should be a uniform potential difference between equipotential lines, and each equipotential line should be labeled.

a.

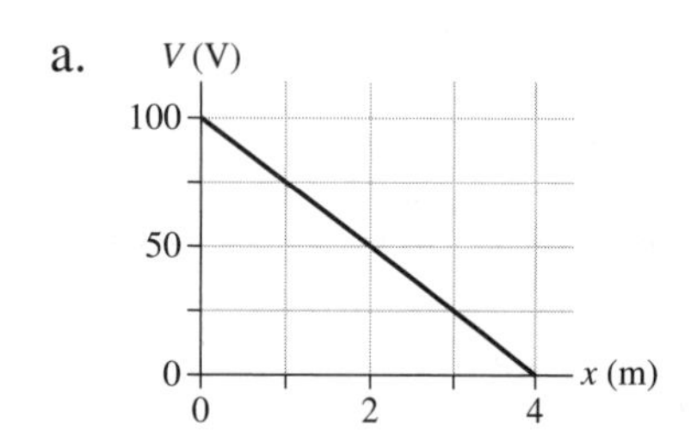

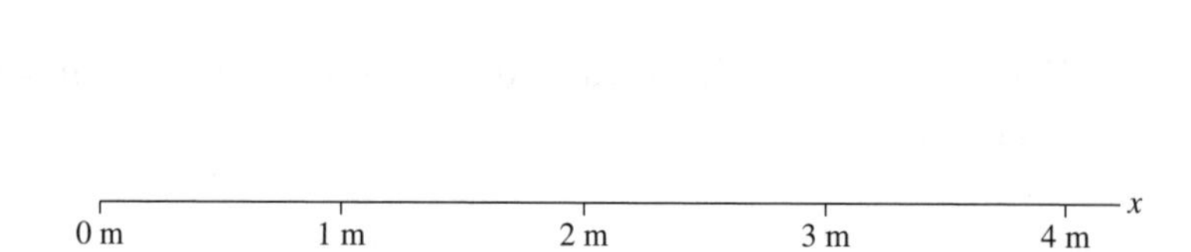

b.

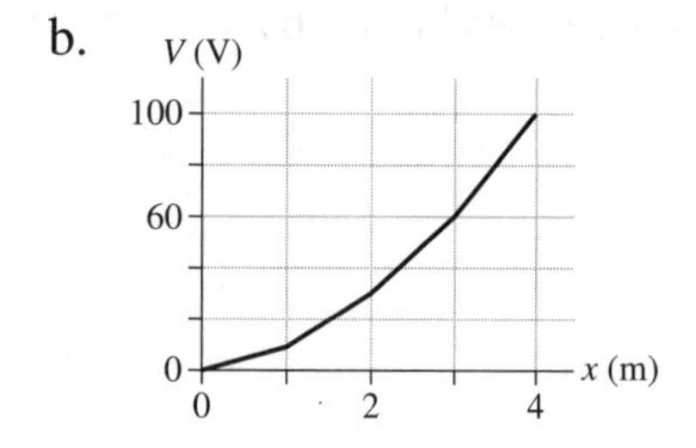

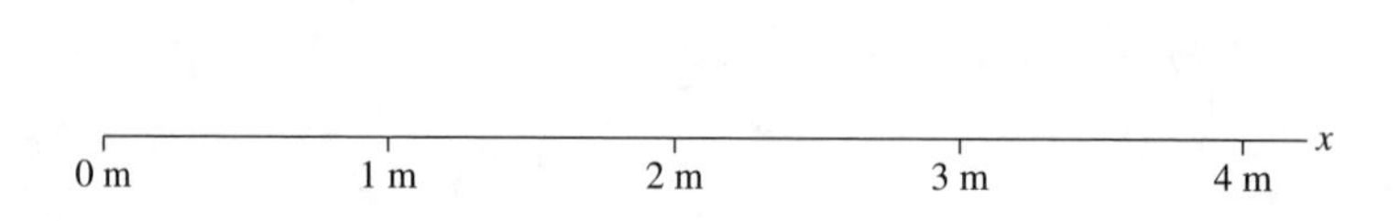

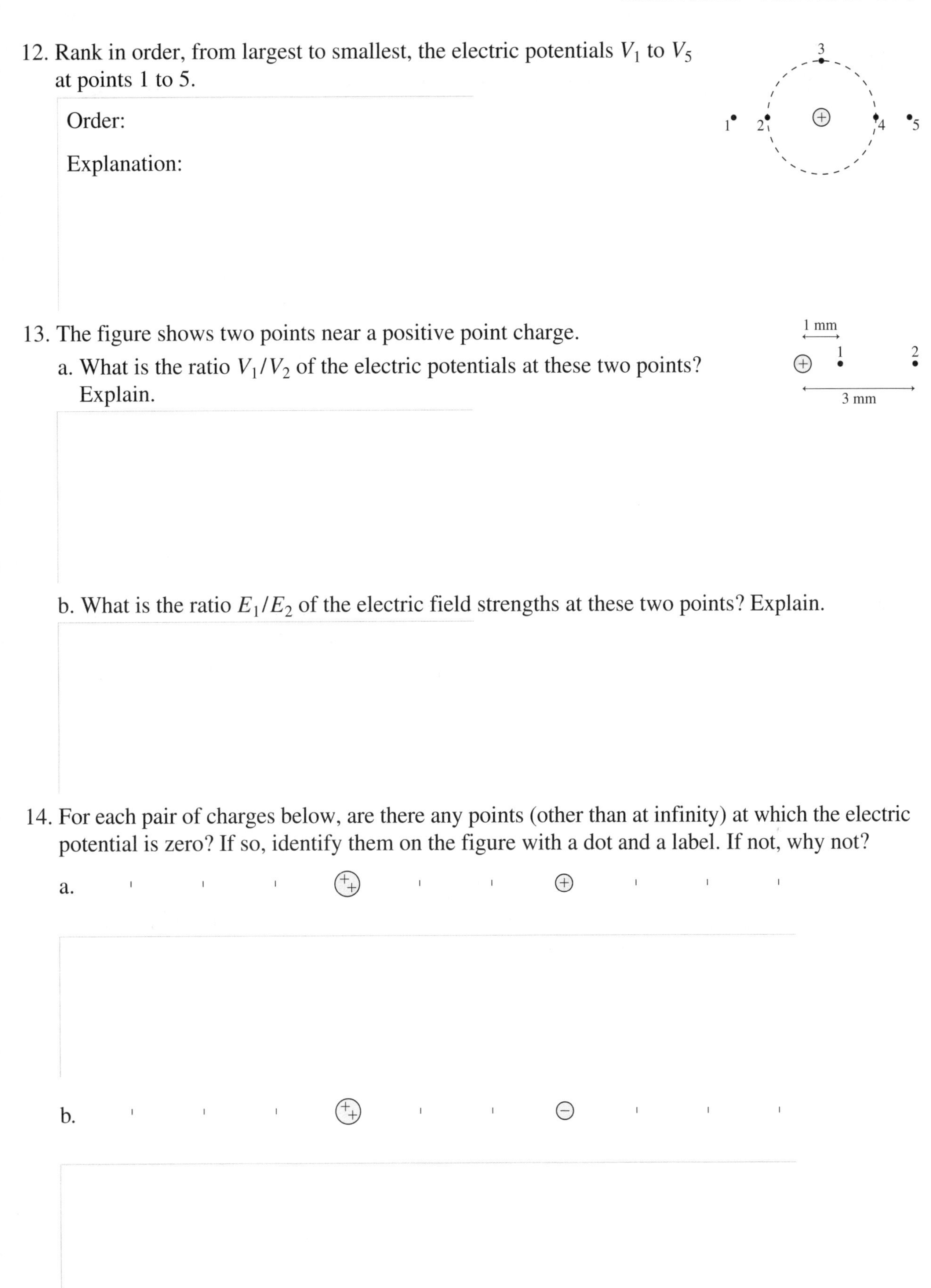

12. Rank in order, from largest to smallest, the electric potentials V_1 to V_5 at points 1 to 5.

Order:

Explanation:

13. The figure shows two points near a positive point charge.

a. What is the ratio V_1/V_2 of the electric potentials at these two points? Explain.

b. What is the ratio E_1/E_2 of the electric field strengths at these two points? Explain.

14. For each pair of charges below, are there any points (other than at infinity) at which the electric potential is zero? If so, identify them on the figure with a dot and a label. If not, why not?

a.

b.

15. An electron moves along the trajectory from i to f.

a. Does the electric potential energy increase, decrease, or stay the same? Explain.

b. Is the electron's speed at f greater than, less than, or equal to its speed at i? Explain.

16. An inflatable metal balloon of radius R is charged to a potential of 1000 V. After all wires and batteries are disconnected, the balloon is inflated to a new radius $2R$.

a. Does the potential of the balloon change as it is inflated? If so, by what factor? If not, why not?

b. Does the potential change at a point distance $r = 4R$ from the center as the balloon is inflated? If so, by what factor? If not, why not?

17. A small charged sphere of radius R_1, mass m_1, and positive charge q_1 is shot head on with speed v_1 from a long distance away toward a second small sphere having radius R_2, mass m_2, and positive charge q_2. The second sphere is held in a fixed location and cannot move. The spheres repel each other, so sphere 1 will slow as it approaches sphere 2. If v_1 is small, sphere 1 will reach a closest point, reverse direction, and be pushed away by sphere 2. If v_1 is large, sphere 1 will crash into sphere 2. For what speed v_1 does sphere 1 just barely touch sphere 2 as it reverses direction?

PSA 21.1

a. Begin by drawing a before-and-after visual overview. Initially, the spheres are far apart and sphere 1 is heading toward sphere 2 with speed v_1. The problem ends with the spheres touching. What is speed of sphere 1 at this instant? How far apart are the centers of the spheres at this instant? Label the before and after pictures with complete information—all in symbolic form.

b. Energy is conserved, so we can use Problem-Solving Approach 21.1. But first we have to identify the "moving charge" q and the "source charge" that creates the potential.

Which is the moving charge? ______ Which is the source charge? ______

c. We're told the charges start "a long distance away" from each other. Based on this statement, what value can you assign to V_i, the potential of the source charge at the initial position of the moving charge? Explain.

d. Now write an expression in terms of the symbols defined above (and any constants that are needed) for the initial energy $K_i + qV_i$.

$K_i + qV_i =$ ______

e. Referring to information on your visual overview, write an expression for the final energy.

$K_f + qV_f =$ ______

f. Energy is conserved, so finish the problem by solving for v_1.

21.5 Connecting Potential and Field

18. For each contour map:
    i. Estimate the electric fields $\vec{E}_a$ and $\vec{E}_b$ at points a and b. Don't forget that $\vec{E}$ is a vector. Show how you made your estimate.
    ii. Draw electric field vectors on top of the contour map.

a.

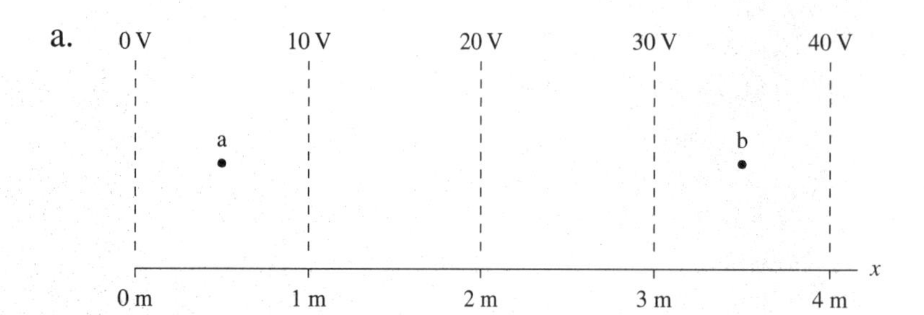

$\vec{E}_a$ = ______

$\vec{E}_b$ = ______

b.

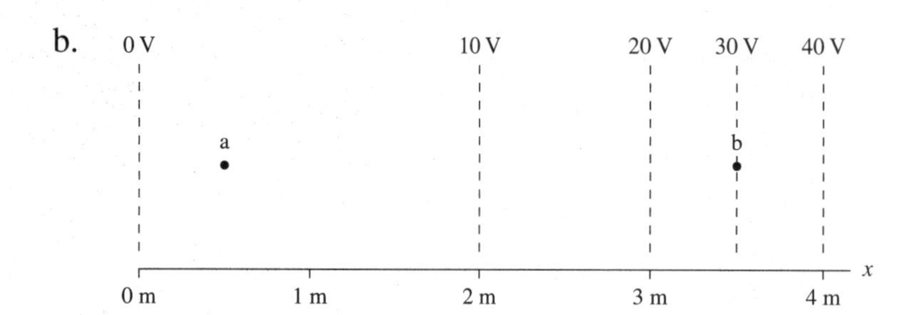

$\vec{E}_a$ = ______

$\vec{E}_b$ = ______

19. Draw the electric field vectors at the dots on this contour map. The length of each vector should be proportional to the field strength at that point.

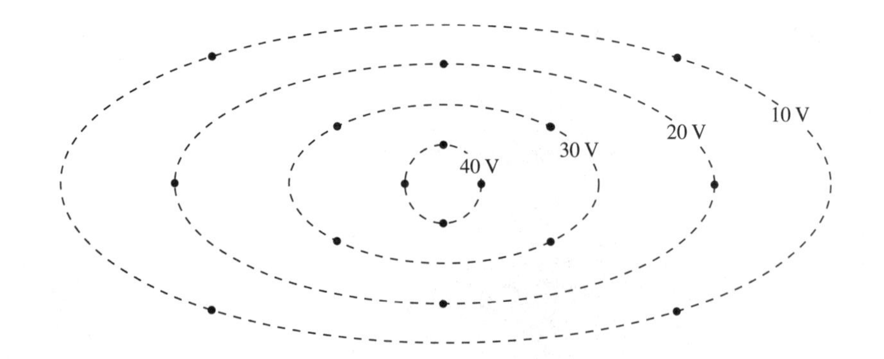

20. Two metal spheres are connected by a metal wire that has a switch in the middle. Initially, the switch is open. Sphere 1, with the larger radius, is given a positive charge. Sphere 2, with the smaller radius, is neutral. Then the switch is closed. Afterward, sphere 1 has charge Q_1, is at potential V_1, and the electric field strength at its surface is E_1. The values for sphere 2 are Q_2, V_2, and E_2.

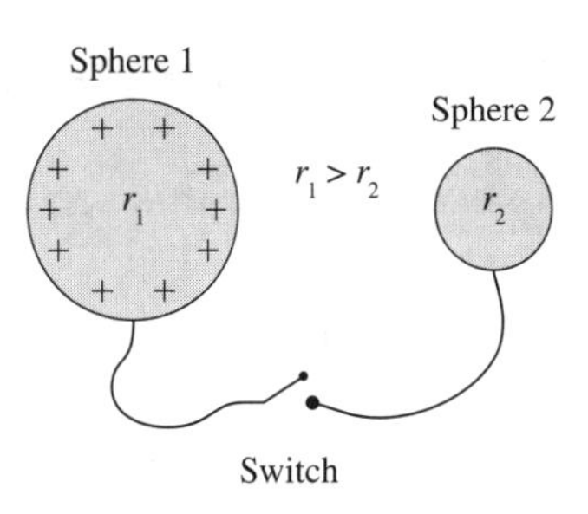

a. Is V_1 larger than, smaller than, or equal to V_2? Explain.

b. Is Q_1 larger than, smaller than, or equal to Q_2? Explain.

c. Is E_1 larger than, smaller than, or equal to E_2? Explain.

21. The figure shows a hollow metal sphere. A negatively charged rod touches the top of the sphere, transferring charge to the sphere. Then the rod is removed.

a. Show on the figure the equilibrium distribution of charge.

b. Draw the electric field diagram.

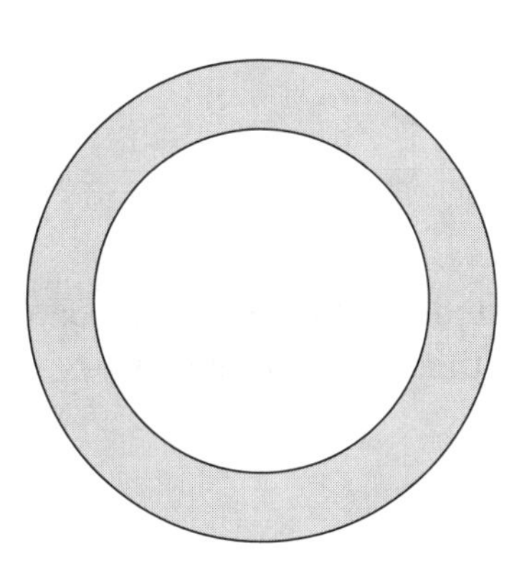

21.6 The Electrocardiogram

No exercises for this section.

21.7 Capacitance and Capacitors

21.8 Energy and Capacitors

22. Rank in order, from largest to smallest, the potential differences $(\Delta V_C)_1$ to $(\Delta V_C)_4$ of these four capacitors.

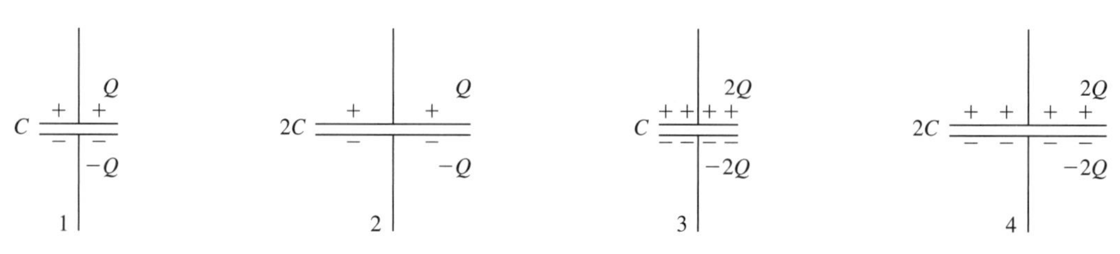

Order:

Explanation:

23. A parallel-plate capacitor has capacitance C. Suppose all three dimensions of the capacitor are doubled—that is, increased by a factor of 2. By what factor does the capacitance increase? Explain.

24. The plates of a parallel-plate capacitor are connected to a battery. If the battery voltage is doubled, by what factor does the energy stored in the capacitor increase?

25. A parallel-plate capacitor is charged, and then disconnected from the battery that charged it; both plates are now electrically isolated. The capacitor charge, capacitance, and potential difference are Q_i, C_i, and $(\Delta V_C)_i$. Then a dielectric is inserted between the plates. Afterward, the charge, capacitance, and potential difference are Q_f, C_f, and $(\Delta V_C)_f$.

    a. Is Q_f larger than, smaller than, or the same as Q_i? Explain.

    b. Is C_f larger than, smaller than, or the same as C_i? Explain.

    c. Is $(\Delta V_C)_f$ larger than, smaller than, or the same as $(\Delta V_C)_i$? Explain.

26. Rank in order, from largest to smallest, the energies $(U_C)_1$ to $(U_C)_4$ stored in each of these capacitors.

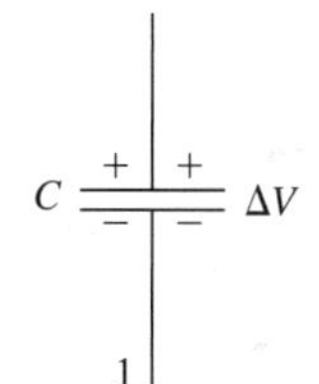

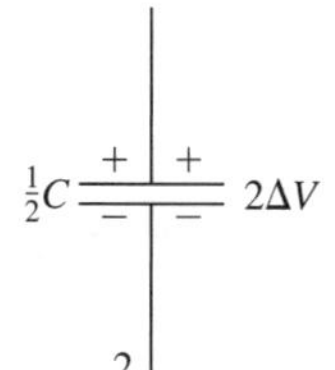

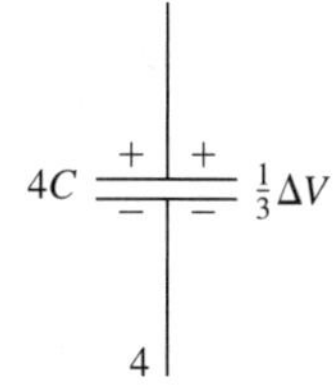

Order:

Explanation:

You Write the Problem!

Exercises 27–30: You are given the equation that is used to solve a problem. For each of these:

a. Write a *realistic* physics problem for which this is the correct equation. Look at worked examples and end-of-chapter problems in the textbook to see what realistic physics problems are like.

b. Finish the solution of the problem.

27. $$\frac{(9.0 \times 10^9\ \text{N} \cdot \text{m}^2/\text{C}^2)\ q_1 q_2}{0.030\ \text{m}} = 9.0 \times 10^{-5}\ \text{J}$$

$$q_1 + q_2 = 40\ \text{nC}$$

28. $$\frac{(9.0 \times 10^9\ \text{N} \cdot \text{m}^2/\text{C}^2)(3.0 \times 10^{-9}\ \text{C})}{r} = 18{,}000\ \text{V}$$

29. $$\tfrac{1}{2}(1.67 \times 10^{-27}\ \text{kg})\, v_\text{i}^2 + 0 = 0 + \frac{(9.0 \times 10^9\ \text{N} \cdot \text{m}^2/\text{C}^2)(2.0 \times 10^{-9}\ \text{C})(1.60 \times 10^{-19}\ \text{C})}{0.0010\ \text{m}}$$

30. $$400\ \text{nC} = (100\ \text{V})\, C$$

$$C = \frac{(8.85 \times 10^{-12}\ \text{F/m})(0.10\ \text{m} \times 0.10\ \text{m})}{d}$$

22 Current and Resistance

22.1 A Model of Current

22.2 Defining and Describing Current

1. a. Describe an experiment that provides evidence that current consists of charge flowing through a conductor. Use both pictures and words.

   b. One model of current is that it consists of the motion of discrete charged particles. Another model is that current is the flow of a continuous charged fluid. Does the experiment you described in part a provide evidence in favor of either one of these models? If so, describe how.

2. Figure A shows capacitor plates that have been charged to $\pm Q$. A very long conducting wire is then connected to the plates as shown in Figure B. Will the separated charges remain on the plates where they are attracted to each other by Coulomb attraction, or will the charges travel through the long wire to discharge the plates? Explain.

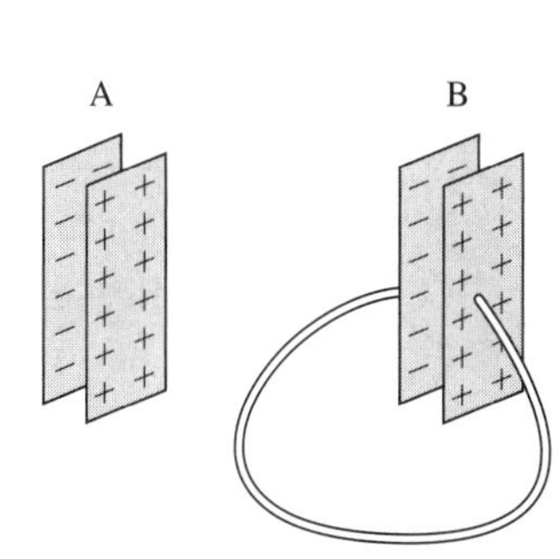

3. A discharging capacitor is used to light bulbs of different types and at different locations as shown.

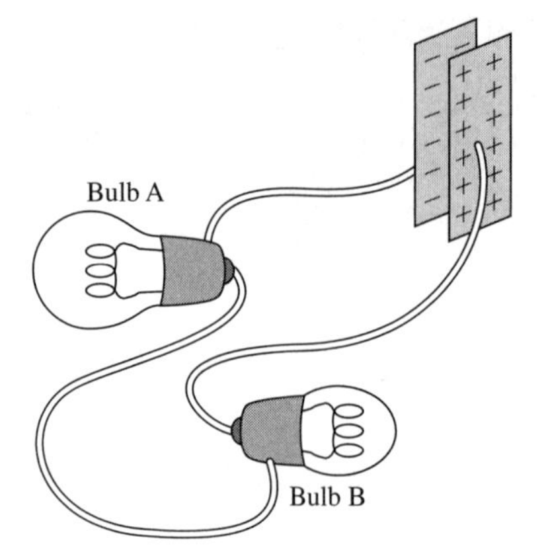

   a. Which of the following, if any, will cause the current in Bulb A to be different than in Bulb B? Explain.
      - Bulb A is rated at a higher wattage than Bulb B.
      - Bulb A is connected directly to the negative plate, while Bulb B is connected to the positive plate.
      - Bulb A is closer (connected through less wire) to the capacitor plates than Bulb B.
      - None of these.

   b. Which of the following, if any, will cause Bulb A to light before Bulb B? Which will cause Bulb B to light before Bulb A? Explain.
      - Bulb A is rated at a higher wattage than Bulb B.
      - Bulb A is connected directly to the negative plate, while Bulb B is connected to the positive plate.
      - Bulb A is closer to the capacitor plates than Bulb B.
      - None of these.

4. Is I_2 greater than, less than, or equal to I_1? Explain.

5. All wires in this figure are made of the same material and have the same diameter. Rank in order, from largest to smallest, the currents I_1 to I_4.

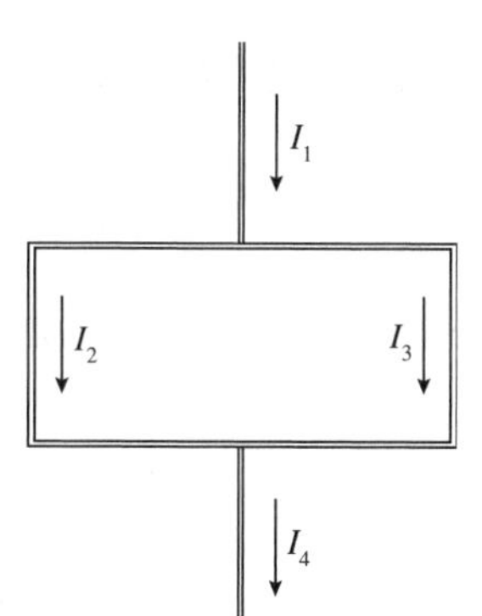

Order:

Explanation:

6. A wire carries a 4 A current. What is the current in the second wire that delivers twice as much charge in half the time?

7. What is the size of the current in the fourth wire? Is the current into or out of the junction? Explain.

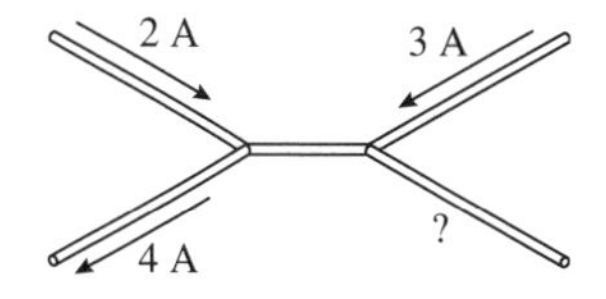

22.3 Batteries and emf

22.4 Connecting Potential and Current

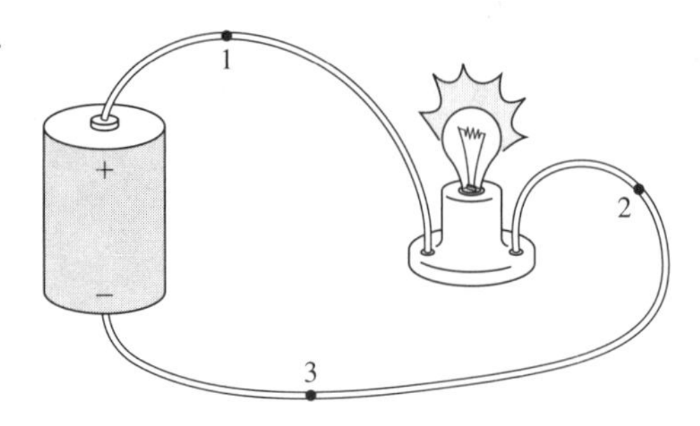

8. A light bulb is connected to a battery with 1-mm-diameter wires. The bulb is glowing.
   a. Draw arrows at points 1, 2, and 3 to show the direction of the electric field at those points. (The points are *inside* the wire.)
   b. Rank in order, from largest to smallest, the currents I_1, I_2, and I_3 at points 1 to 3.

Order:

Explanation:

9. You have a long wire with resistance R. You would like to have a wire of the same length but resistance $2R$. Should you (a) change to a wire of the same diameter but made of a material having twice the resistivity, or (b) change to a wire made of the same material but with half the diameter? Or will either do? Explain.

10. Wire 1 and wire 2 are made of the same metal and are the same length. Wire 1 has twice the diameter and half the potential difference across its ends. What is the ratio of I_1/I_2?

22.5 Ohm's Law and Resistor Circuits

11. A wire consists of two equal-diameter segments. Their resistivities differ, with $\rho_1 > \rho_2$. The current in segment 1 is I_1. Compare the values of the currents in the two segments. Is I_2 greater than, less than, or equal to I_1? Explain.

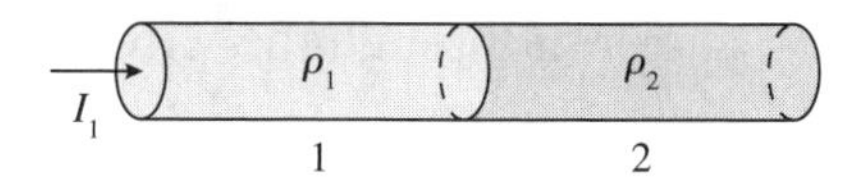

12. A graph of current as a function of potential difference is given for a particular wire segment.

    a. What is the resistance of the wire?

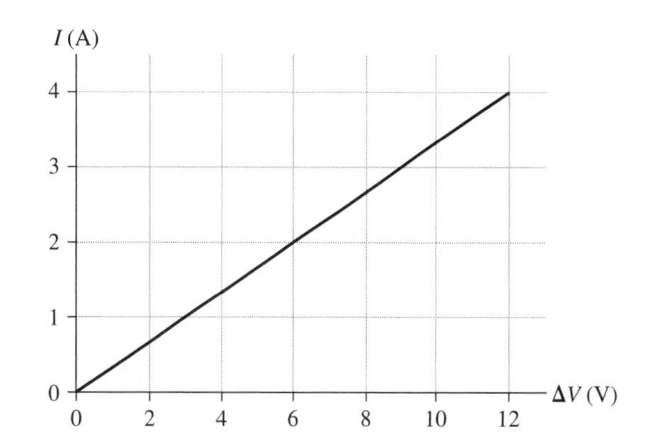

   b. Sketch and label on the same axes the graph of current vs. potential difference for a wire made of the same material but twice as long as the wire in part a.

   c. Sketch and label on the same axes the graph of current vs. potential difference for a wire made of the same material but with twice the cross-section area of the wire in part a.

13. For resistors R_1 and R_2:

    a. Which end (left, right, top, or bottom) is more positive?

    R_1: ______________ R_2: ______________

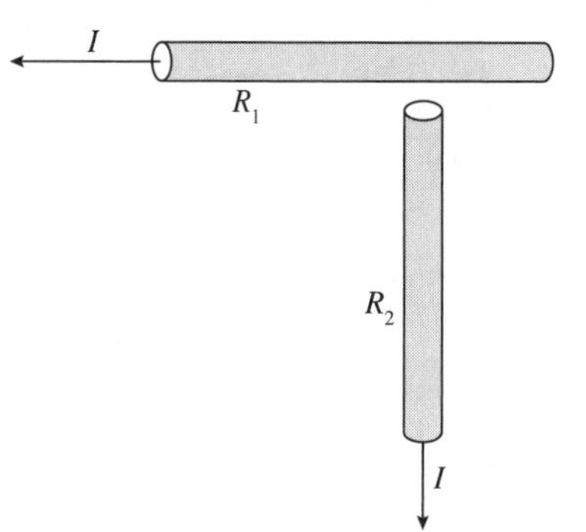

   b. In which direction (such as left to right or top to bottom) does the potential decrease?

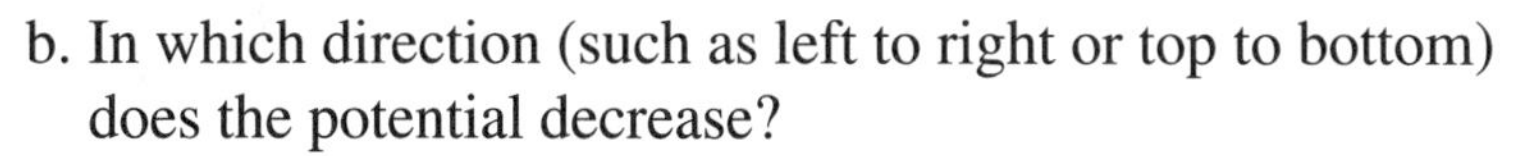

   R_1: ______________

   R_2: ______________

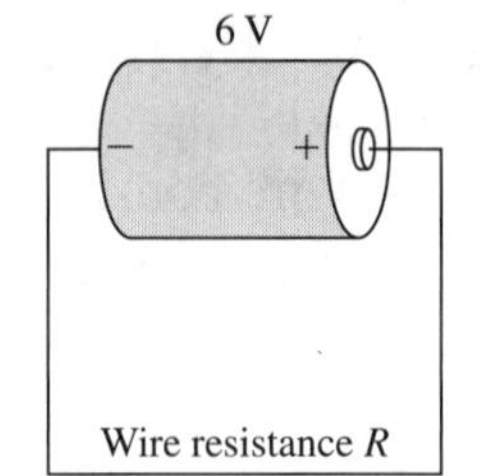

14. A wire is connected to the terminals of a 6 V battery. What is the potential difference ΔV_{wire} between the ends of the wire, and what is the current I through the wire, if the wire has the following resistances:

a. $R = 1\ \Omega$ $\Delta V_{\text{wire}} =$ ________ $I =$ ________

b. $R = 2\ \Omega$ $\Delta V_{\text{wire}} =$ ________ $I =$ ________

c. $R = 3\ \Omega$ $\Delta V_{\text{wire}} =$ ________ $I =$ ________

d. $R = 6\ \Omega$ $\Delta V_{\text{wire}} =$ ________ $I =$ ________

15. Rank in order, from largest to smallest, the currents I_1 to I_4 through these four resistors.

+ 2 V −	+ 1 V −	+ 2 V −	+ 1 V −
2 Ω, I_1	2 Ω, I_2	1 Ω, I_3	1 Ω, I_4

Order:

Explanation:

22.6 Energy and Power

16. a. Two conductors of equal lengths are connected in a line so that the same current I flows through both. The conductors are made of the same material but have different radii. Which of the two conductors dissipates the larger amount of power? Explain.

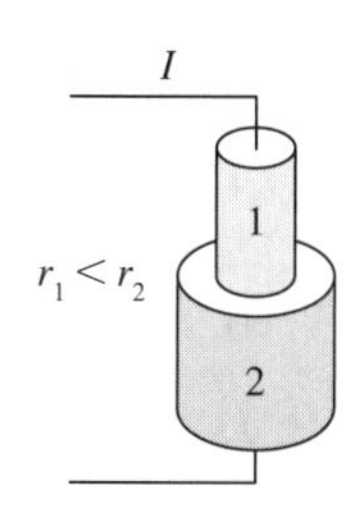

b. The same two conductors in part a are then each connected separately to identical batteries so that each has the same potential difference across it. Which of the two conductors now dissipates the larger amount of power? Explain.

17. Two resistors of equal lengths are connected to a battery by ideal wires. The resistors have the same radii but are made of different materials and have different resistivities ρ with $\rho_1 > \rho_2$.

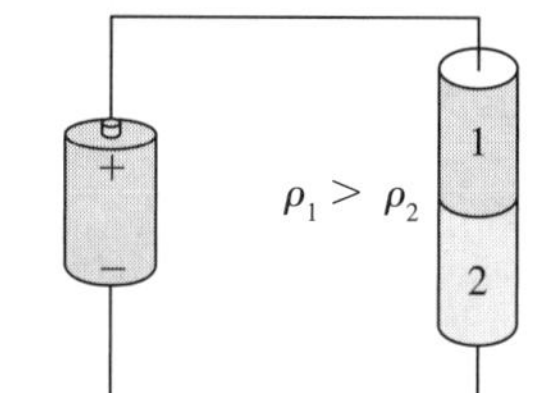

a. Is the current I_1 in resistor 1 larger than, smaller than, or the same as I_2 in resistor 2? Explain.

b. Which of the two resistors dissipates the larger amount of power? Explain.

c. Is the voltage ΔV_1 across resistor 1 larger than, smaller than, or the same as ΔV_2 across resistor 2? Explain.

You Write the Problem!

Exercises 18–20: You are given the equation that is used to solve a problem. For each of these:

a. Write a *realistic* physics problem for which this is the correct equation. Look at worked examples and end-of-chapter problems in the textbook to see what realistic physics problems are like.
b. Finish the solution of the problem.

18. $I \times (150\ \Omega) = 4.5\ \text{V}$

19. $100\ \text{W} = \dfrac{(120\ \text{V})^2}{R}$

20. $(2.0\ \text{A}) \times \dfrac{(1.5 \times 10^{-6}\ \Omega \cdot \text{m})L}{\pi(7.5 \times 10^{-4}\ \text{m})^2} = 12\ \text{V}$

23 Circuits

23.1 Circuit Elements and Diagrams

23.2 Kirchhoff's Laws

Exercises 1–4: Redraw the circuits shown using standard circuit symbols with only right angle corners.

1.

2.

3.

4.

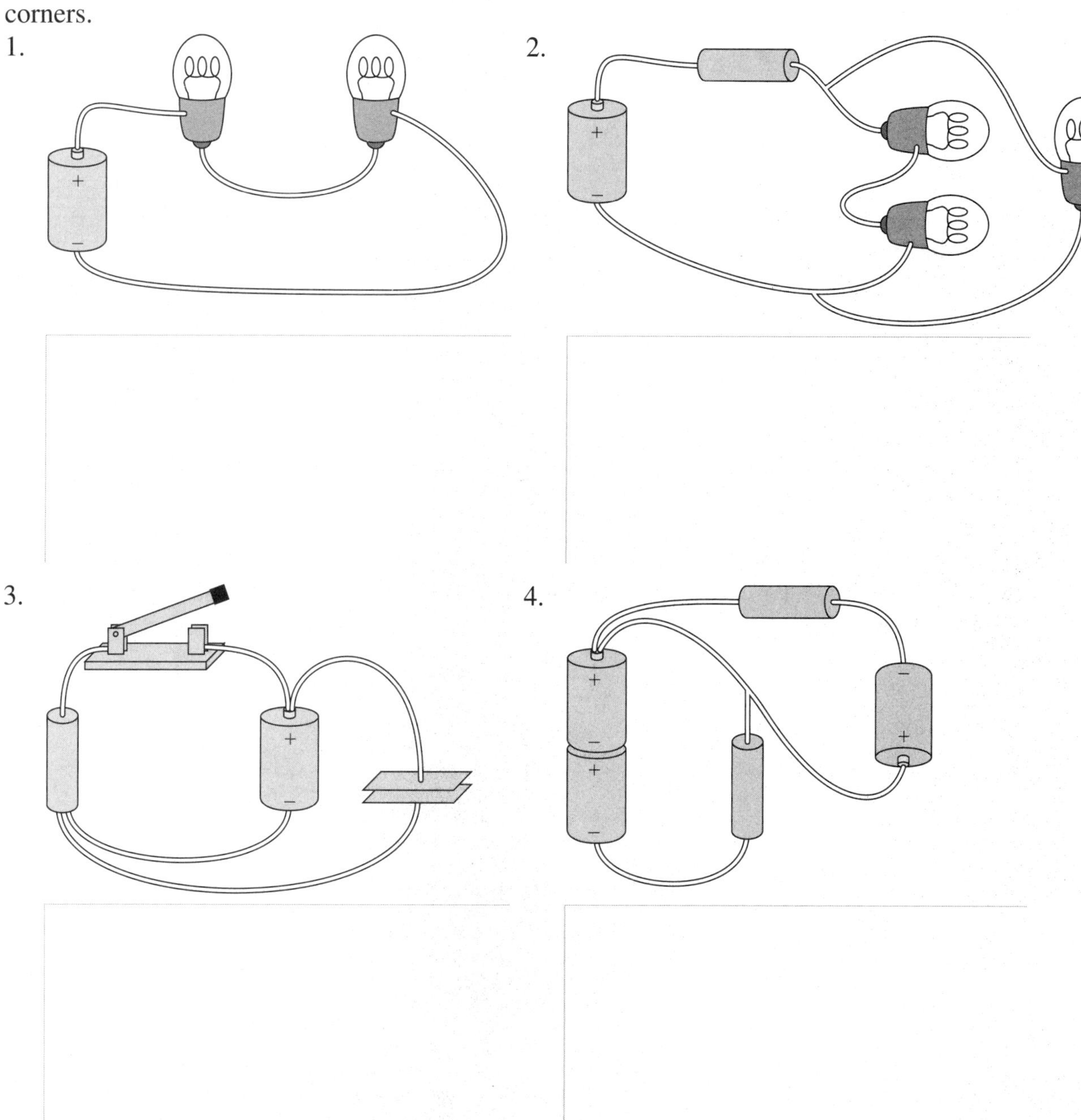

5. The tip of a flashlight bulb is touching the top of a 3 V battery. Does the bulb light? Why or why not?

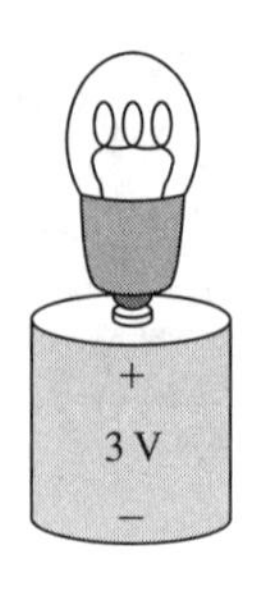

6. A flashlight bulb is connected between two 1.5 V batteries as shown. Does the bulb light? Why or why not?

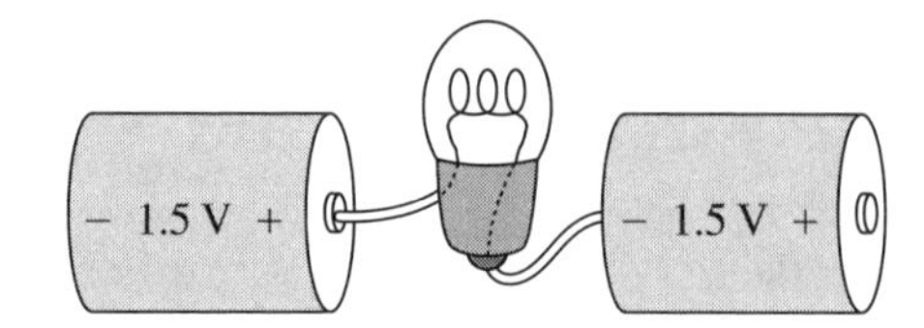

7. Current I_{in} flows into three resistors connected together one after the other. The graph shows the value of the potential as a function of distance.

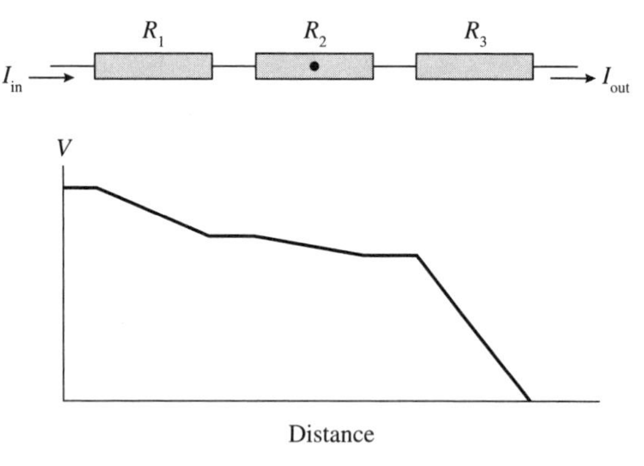

a. Is I_{out} greater than, less than, or equal to I_{in}? Explain.

b. Rank in order, from largest to smallest, the three resistances R_1, R_2, and R_3.

Order:

Explanation:

c. Is there an electric field at the point inside R_2 that is marked with a dot? If so, in which direction does it point? If not, why not?

8. a. Redraw the circuit shown as a standard circuit diagram.

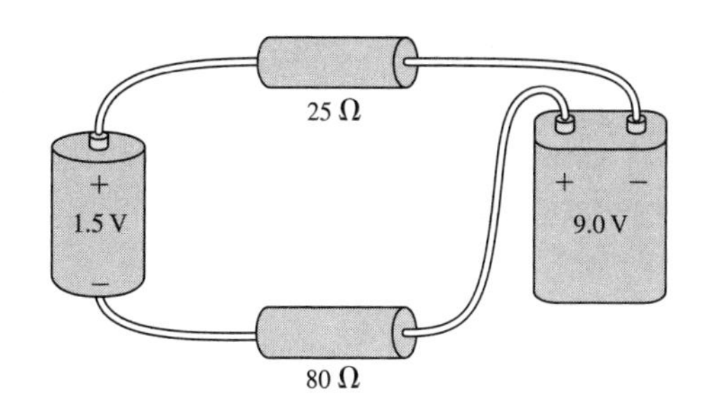

b. Assign a direction for the current and label the current arrow I on your sketch.
c. Apply Kirchhoff's loop law to determine the current through the resistors.

9. Draw a circuit for which the Kirchhoff loop law equation is

$$6\text{V} - I\cdot 2\Omega + 3\text{V} - I\cdot 4\Omega = 0$$

Assume that the analysis is done in a clockwise direction.

10. The current in a circuit is 2.0 A. The graph shows how the potential changes when going around the circuit in a clockwise direction, starting from the lower left corner. Draw the circuit diagram.

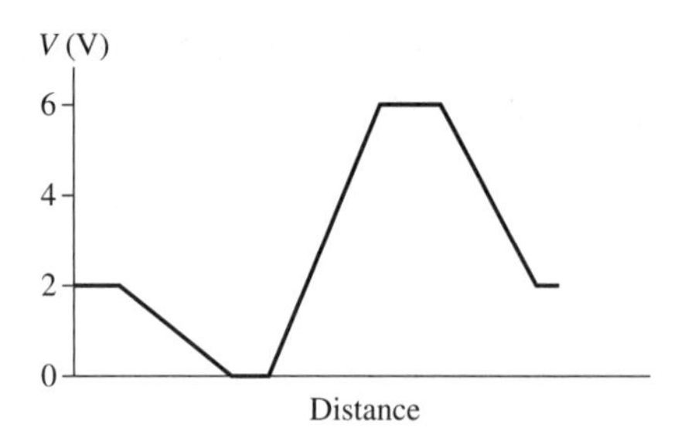

23.3 Series and Parallel Circuits

11. What is the equivalent resistance of each group of resistors?

a.

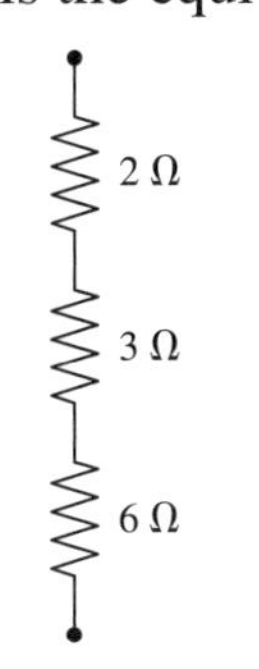

$R_{eq} =$ ______

b.

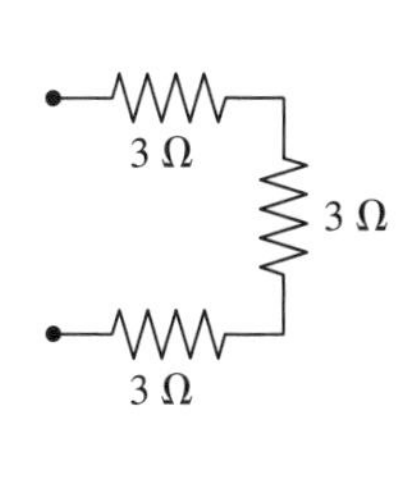

$R_{eq} =$ ______

c.

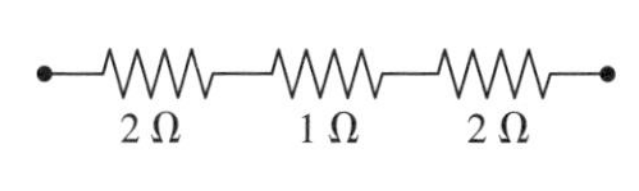

$R_{eq} =$ ______

12. What is the equivalent resistance of each group of resistors?

a.

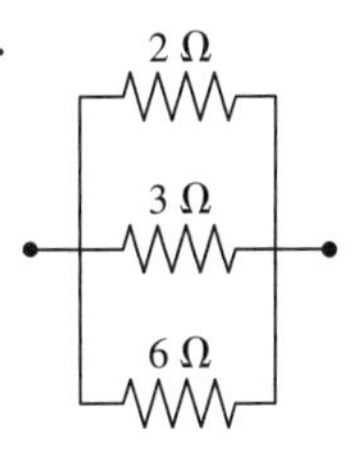

$R_{eq} =$ ______

b.

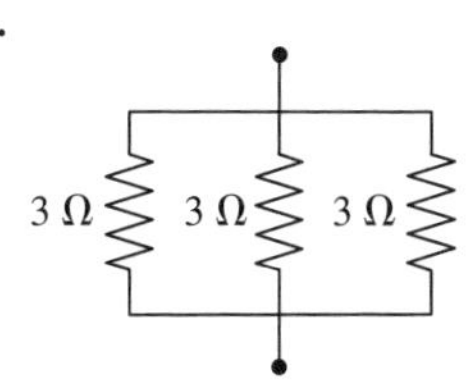

$R_{eq} =$ ______

c.

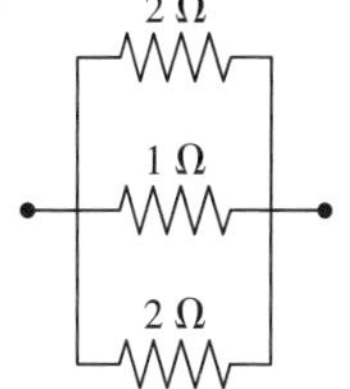

$R_{eq} =$ ______

13. The figure shows five combinations of identical resistors. Rank in order, from largest to smallest, the equivalent resistances $(R_{eq})_1$ to $(R_{eq})_5$.

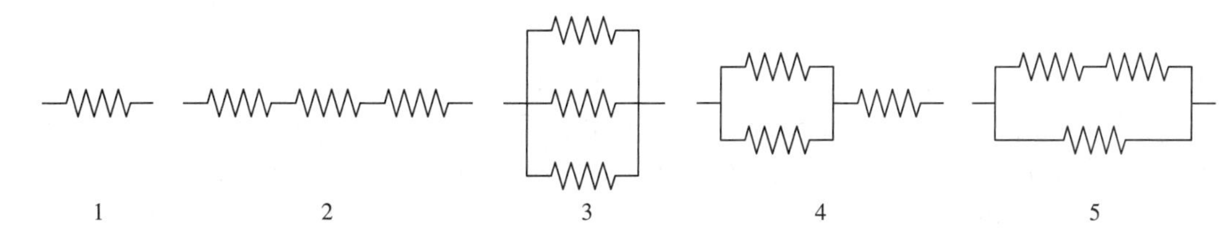

Order:

Explanation:

14. A 60 W light bulb and a 100 W light bulb are placed one after the other in a circuit. The battery's emf is large enough that both bulbs are glowing. Which one glows more brightly? Explain.

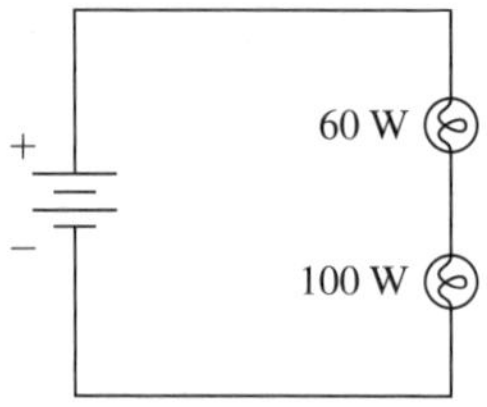

60 W

100 W

15. In your own words, without using a mathematical formula, state why the equivalent resistance of any number of resistors in series is always greater than any of the individual resistances and why the equivalent resistance of any number of resistances in parallel is always less than any of the individual resistances.

16. Are the three resistors shown wired in series, parallel, or a combination of series and parallel? To test your conclusion, trace all connections to the positive terminal of the battery in red. (These are all at the same potential.) Now trace all connections to the negative side of the battery in blue. (These are also at a common potential.) What do your tracings imply about how these three resistors are wired?

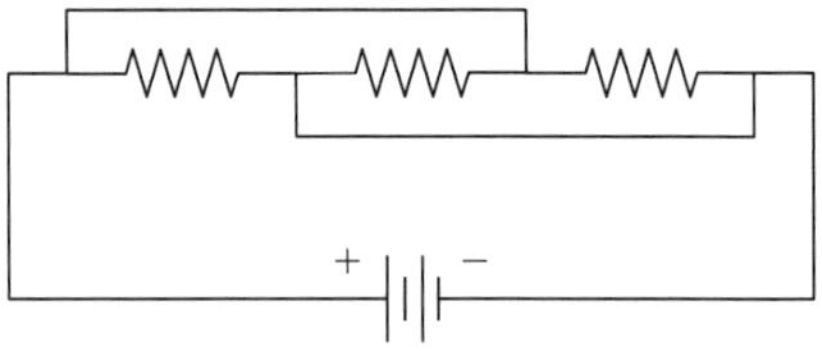

23.4 Measuring Voltage and Current

17. A student assigned the task of measuring the current through the 50 Ω resistor sets up the circuit shown.

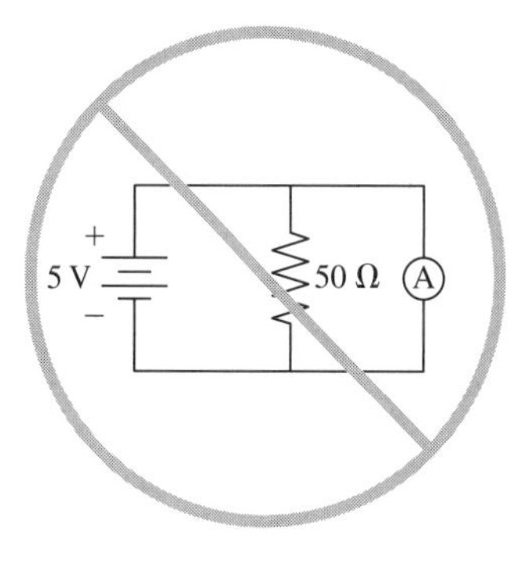

a. Why is this method wrong?

b. Draw a circuit to show how the current *should* be measured.

18. A student assigned the task of measuring the voltage across the 2 Ω resistor sets up the circuit shown.

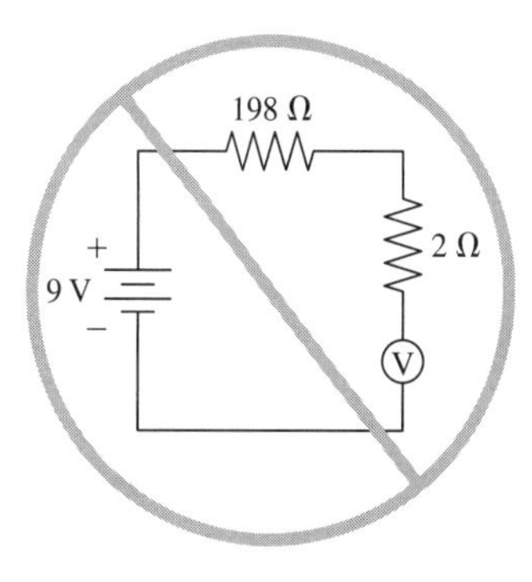

a. Why is this method wrong?

b. Draw a circuit to show how the voltage *should* be measured.

23.5 More Complex Circuits

19. Bulbs A, B, and C are identical. Rank in order, from most to least, the brightness of the three bulbs.

Order:

Explanation:

20. Initially bulbs A and B are glowing. Then the switch is closed. What happens to each bulb? Does it get brighter, stay the same, get dimmer, or go out? Explain your reasoning.

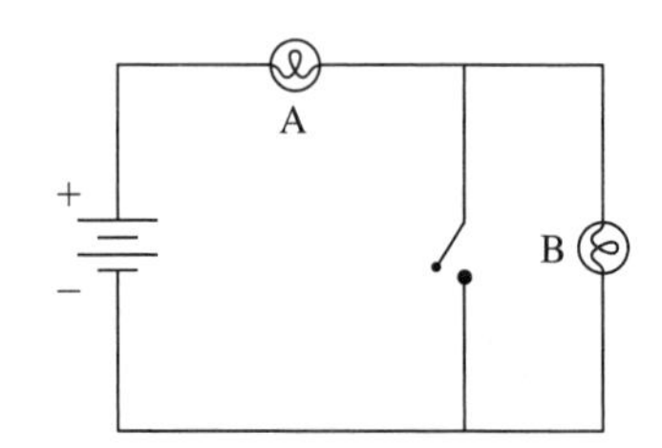

21. Bulbs A and B are identical. Initially, both are glowing.

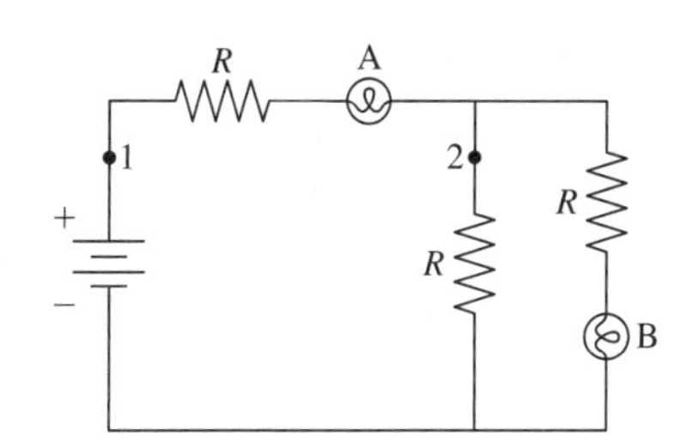

a. Bulb A is removed from its socket. What happens to bulb B? Does it get brighter, stay the same, get dimmer, or go out? Explain.

b. Bulb A is replaced. Bulb B is then removed from its socket. What happens to bulb A? Does it get brighter, stay the same, get dimmer, or go out? Explain.

c. The circuit is restored to its initial condition. A wire is then connected between points 1 and 2. What happens to the brightness of each bulb?

22. Two batteries are identical and the four resistors all have exactly the same resistance.

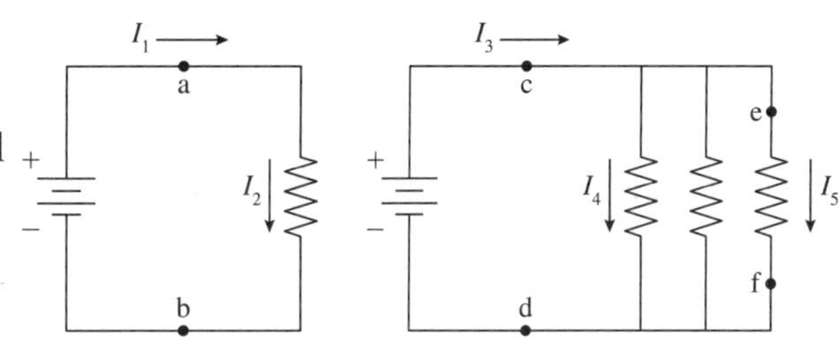

a. Compare ΔV_{ab}, ΔV_{cd}, and ΔV_{ef}. Are they all the same? If not, rank them in decreasing order. Explain your reasoning.

b. Rank in order, from largest to smallest, the five currents I_1 to I_5.

Order:

Explanation:

23. Real circuits frequently have a need to reduce a voltage to a smaller value. This is done with a *voltage-divider circuit* such as the one shown here. We've shown the input voltage V_{in} as a battery, but in practice, it might be a voltage signal produced by some other circuit, such as the receiver circuit in your television.

PSA 23.1

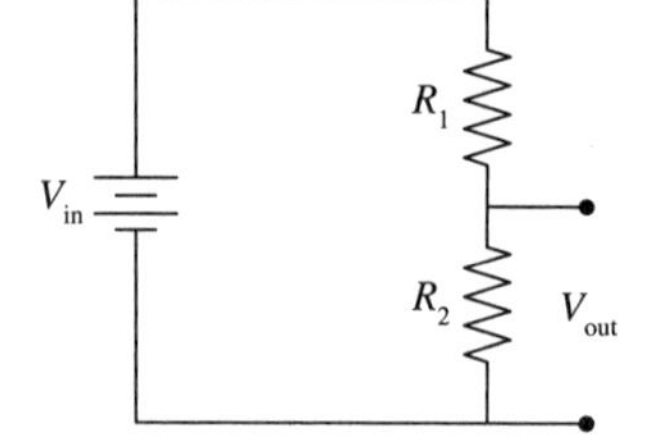

a. Is this a series or a parallel circuit? ____________

b. Symbolically, what is the value of the equivalent resistance? R_{eq} = ____________

c. Redraw the circuit with the two resistors replaced with the equivalent resistance. Note that V_{out} will not be seen in this circuit.

d. You now have one resistor connected directly across the battery. Find the voltage across and the current through this resistor. Write your answers in terms of V_{in}, R_1, and R_2.

$\Delta V =$ ____________ $I =$ ____________

e. Which of these, ΔV or I, is the same for R_1 and R_2 as for R_{eq}? ____________

f. Using Ohm's law and your answer to e, what is V_{out}? It's most useful to write your answer as $V_{out} = \text{something} \times V_{in}$.

$V_{out} =$ ____________

g. What is V_{out} if $R_2 = 2R_1$? $V_{out} =$ ____________

23.6 Capacitors in Parallel and Series

24. Each capacitor in the circuits below has capacitance C. What is the equivalent capacitance of the group of capacitors?

a.

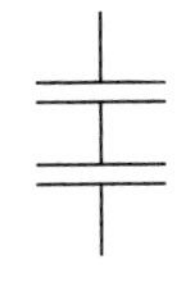

$C_{eq} =$ ______

b.

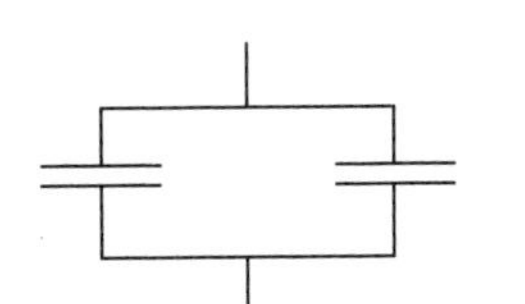

$C_{eq} =$ ______

c.

$C_{eq} =$ ______

d.

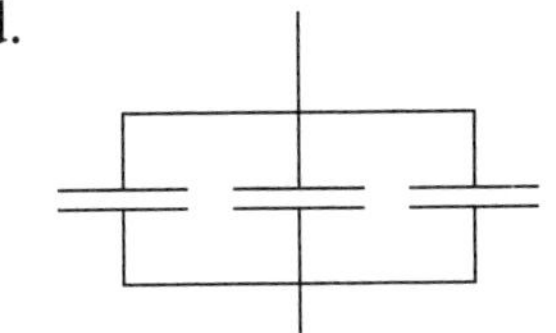

$C_{eq} =$ ______

e.

$C_{eq} =$ ______

f.

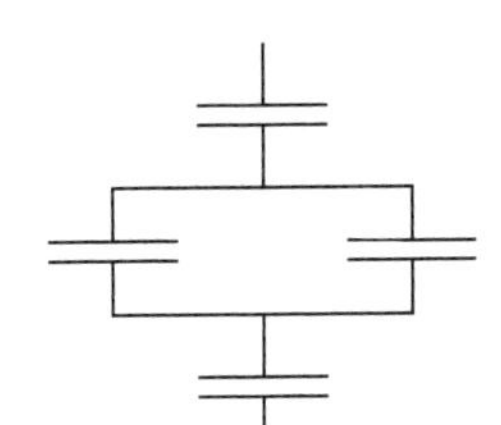

$C_{eq} =$ ______

25. Rank in order, from largest to smallest, the equivalent capacitances $(C_{eq})_1$ to $(C_{eq})_4$ of these four groups of capacitors.

C C C

1

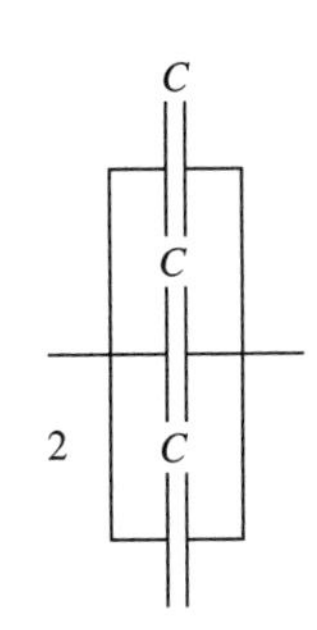

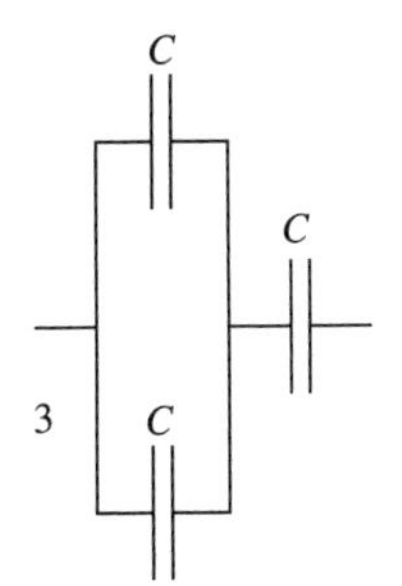

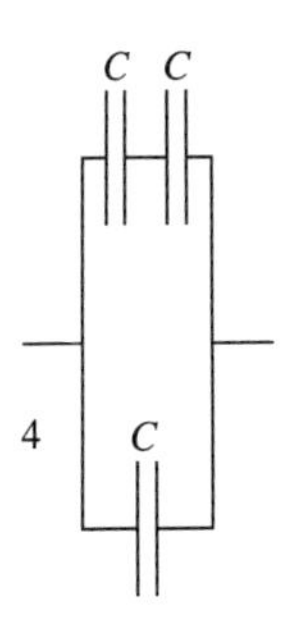

Order:

Explanation:

23.7 *RC* Circuits

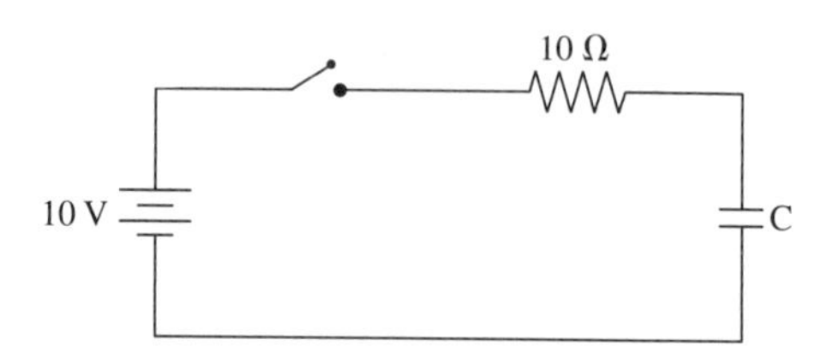

26. The charge on the capacitor is zero when the switch closes at $t = 0$ s.

    a. What will be the current in the circuit after the switch has been closed for a long time? Explain.

    b. Immediately after the switch closes, before the capacitor has had time to charge, the potential difference across the capacitor is zero. What must be the potential difference across the resistor in order to satisfy Kirchhoff's loop law? Explain.

    c. Based on your answer to part b, what is the current in the circuit immediately after the switch closes?

    d. Sketch a graph of current versus time, starting from just before $t = 0$ s and continuing until the switch has been closed a long time. There are no numerical values for the horizontal axis, so you should think about the *shape* of the graph.

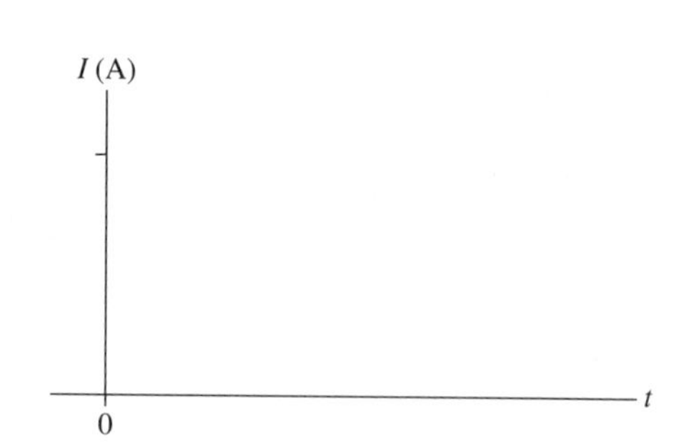

27. Graphs below show voltage versus time for the capacitor and the resistor in the *RC* circuit at the right.

    a. Write the appropriate times in the blanks below the horizontal axes of the first two graphs. Don't forget to include units.

    b. Complete the third graph by showing the *sum* of the voltages across the resistor and the capacitor and by writing the appropriate times in the blanks below the axis.

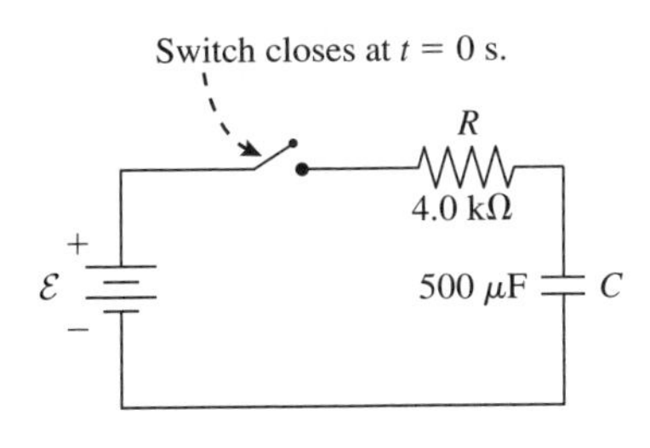

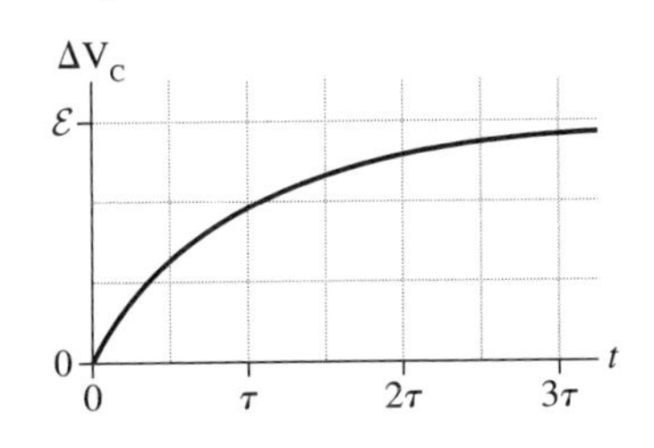

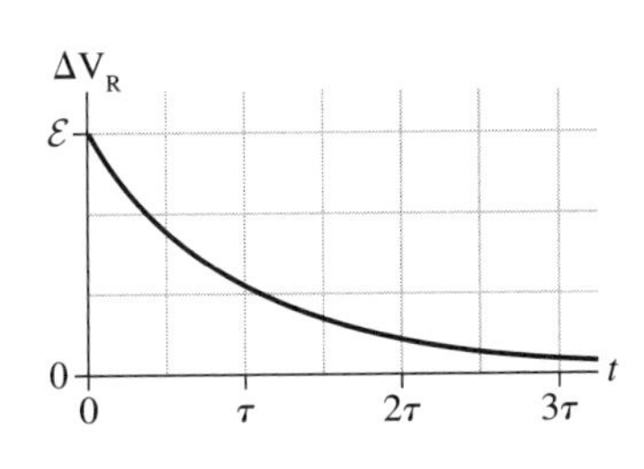

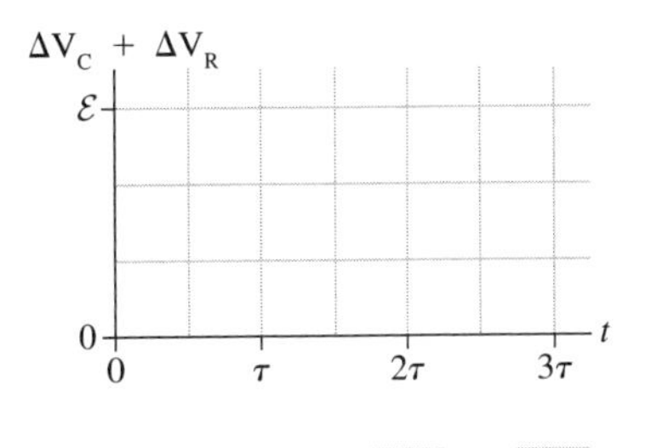

28. The capacitors in each circuit are discharged when the switch closes at $t = 0$ s. Rank in order, from largest to smallest, the time constants τ_1 to τ_5 with which each circuit will discharge.

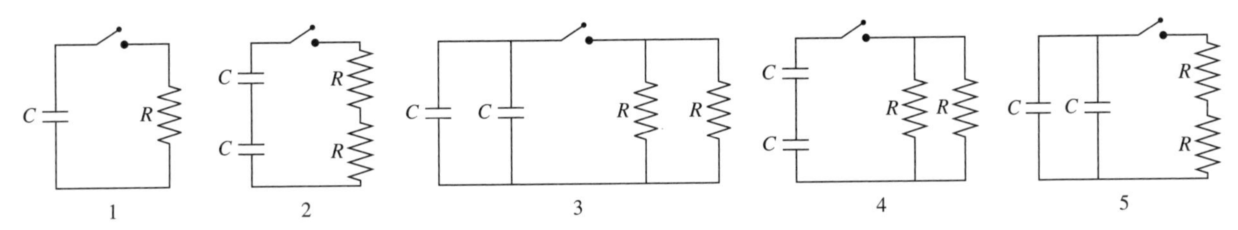

Order:

Explanation:

23.8 Electricity in the Nervous System

No exercises for this section.

24 Magnetic Fields and Forces

24.1 Magnetism

1. The compass needle below is free to rotate in the plane of the page. Either a bar magnet or a charged rod is brought toward the *center* of the compass. Does the compass rotate? If so, does it rotate clockwise or counterclockwise? If not, why not?

a.

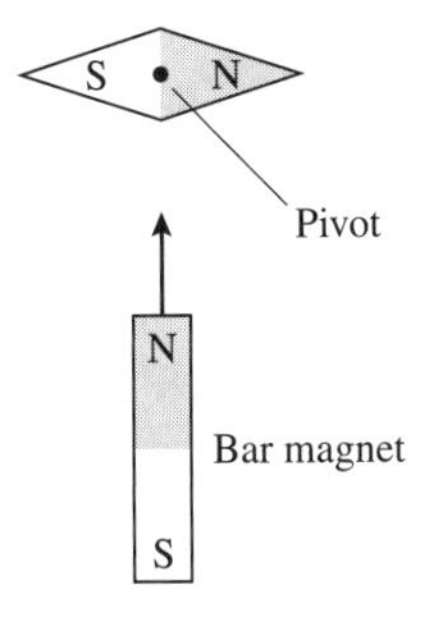

b.

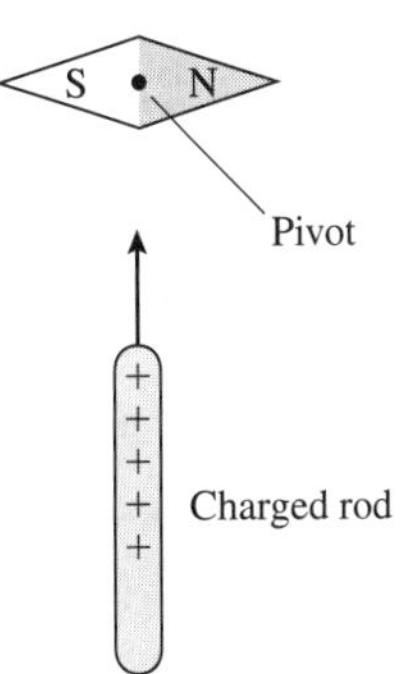

2. You have two electrically neutral metal cylinders that exert strong attractive forces on each other. You have no other metal objects. Can you determine if *both* of the cylinders are magnets, or if one is a magnet and the other just a piece of iron? If so, how? If not, why not?

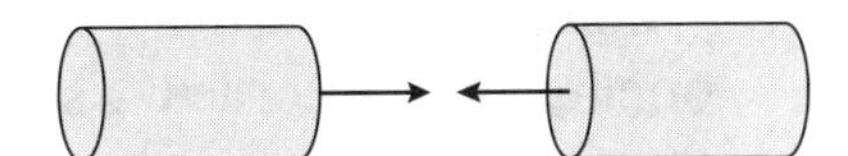

3. Can you think of any kind of object that is repelled by *both* ends of a bar magnet? If so, what? If not, what prevents this from happening?

4. A metal sphere hangs by a thread. When the north pole of a bar magnet is brought near, the sphere is strongly attracted to the magnet. Then the magnet is reversed and its south pole is brought near the sphere. How does the sphere respond? Explain.

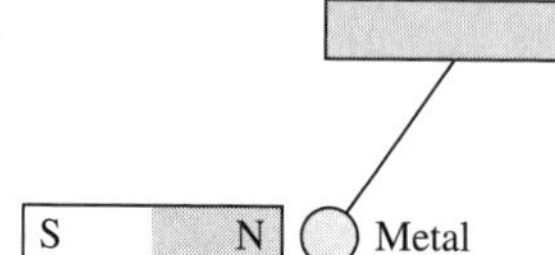

24.2 The Magnetic Field

5. A compass is placed at 12 different positions and its orientation is recorded. Use this information to draw the magnetic *field lines* in this region of space. Draw the field lines on the figure.

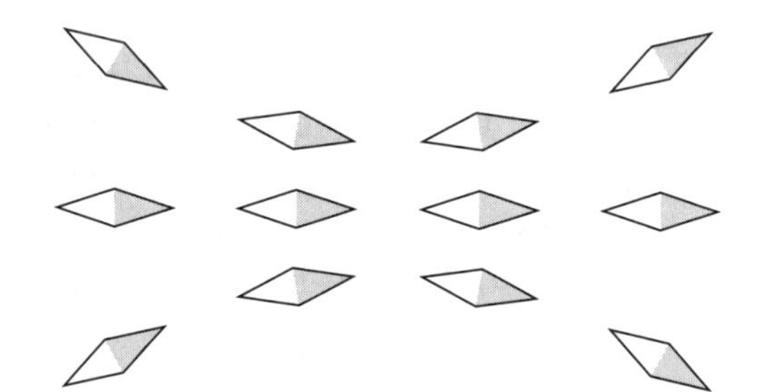

24.3 Electric Currents Also Create Magnetic Fields

6. A neutral copper rod, a polarized insulator rod, and a bar magnet are arranged around a current-carrying wire as shown. For each, will it stay where it is? Move toward or away from the wire? Rotate clockwise or counterclockwise? Explain.

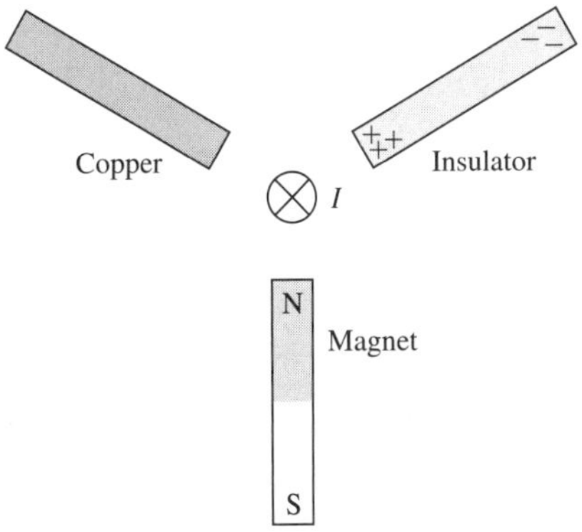

Copper

Insulator

I

N

Magnet

S

a. Neutral copper rod:

b. Insulating rod:

c. Bar magnet:

7. For each of the current-carrying wires shown, draw a compass needle in its equilibrium orientation at the positions of the dots. Label the poles of the compass needle.

a.

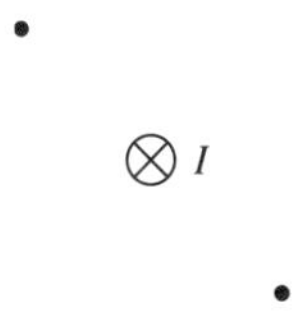

I

b.

I

8. The figure shows a current-carrying wire directed into the page and a nearby compass needle. Is the wire's current going into the page or coming out of the page? Explain.

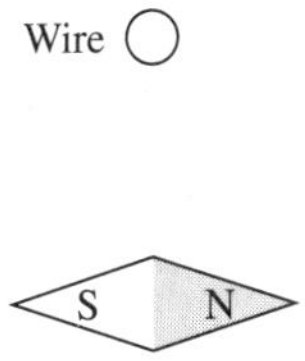

Wire

S N

9. Each figure below shows a current-carrying wire. Draw the magnetic field.

a.

I

The wire is perpendicular to the page. Draw the magnetic field lines. Don't forget to include arrows to show the field direction.

b.

I

The wire is in the plane of the page. Use crosses and dots to show the magnetic field above and below the wire.

10. This current-carrying wire is in the plane of the page. Show the magnetic field on both sides of the wire.

11. Use an arrow to show the current direction in this wire.

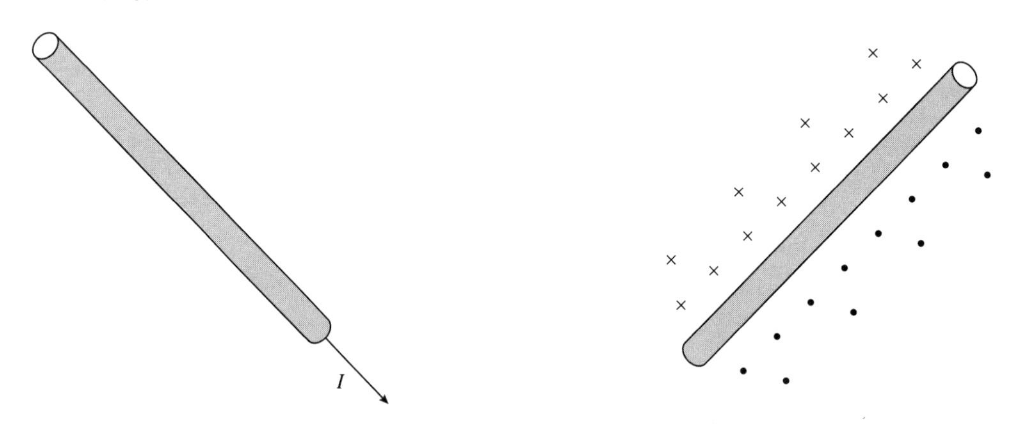

12. Each figure below shows two long straight wires carrying equal currents in and out of the page. At each of the dots, use a **black** pen or pencil to show and label the magnetic fields $\vec{B}_1$ and $\vec{B}_2$ of each wire. Then use a **red** pen or pencil to show the net magnetic field.

a.

Wire 1

Wire 2

b.

Wire 1

Wire 2

24.4 Calculating the Magnetic Field Due to a Current

13. A long straight wire, perpendicular to the page, passes through a uniform magnetic field. The *net* magnetic field at point 3 is zero.
    a. On the figure, show the direction of the current in the wire.
    b. Points 1 and 2 are the same distance from the wire as point 3; point 4 is twice as distant. Construct vector diagrams at points 1, 2, and 4 to determine the net magnetic field at each point.

Uniform
B-field
1
2
Wire
3
4

14. The figure shows the magnetic field seen when facing a current loop in the plane of the page. On the figure, show the direction of the current in the loop.

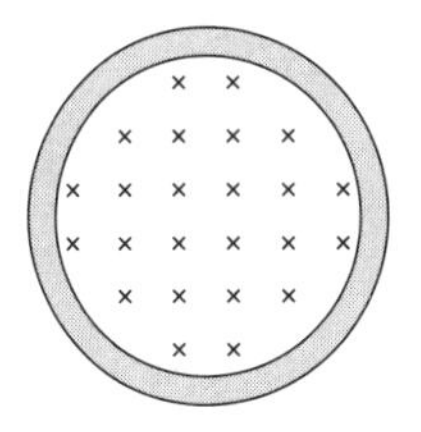

15. Rank in order, from largest to smallest, the magnetic field strengths B_1 to B_3 produced by these three solenoids.

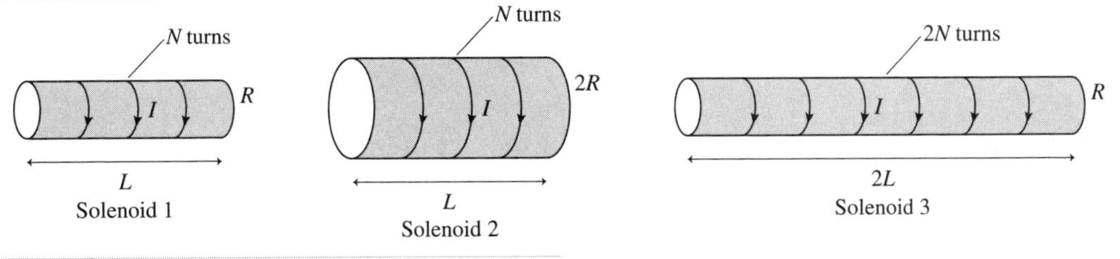

Order:

Explanation:

16. Suppose you need to find the magnetic field near the intersection of two long, straight, current-carrying wires. Assume that one wire lies directly on top of the other. Let the intersection of the wires be the origin of a coordinate system and let the point of interest, which is in the same plane, have coordinates (x, y). Recall that the magnetic field is a vector, having both a magnitude and a direction.

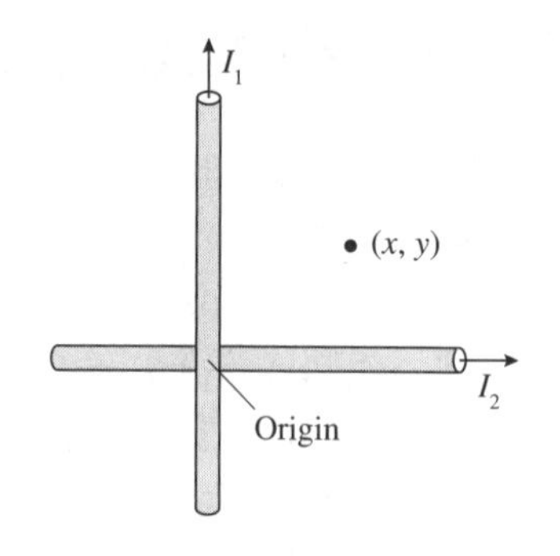

a. At the point of interest, what is the direction of magnetic field $\vec{B}_1$ due to current I_1? Explain.

b. Write an expression for the magnitude of $\vec{B}_1$.

c. What is the direction of magnetic field $\vec{B}_2$ due to current I_2? Explain.

d. Write an expression for the magnitude of $\vec{B}_2$.

e. What are the only two possible directions for the net magnetic field at this point?

f. Would knowing $I_1 > I_2$ be enough information to determine the direction of the net magnetic field? Why or why not?

g. Let a magnetic field pointing out of the page have a positive value, one pointing into the page a negative value. This is actually B_z, the magnetic field along the z-axis. Use your results for parts a–d to write an expression for B_z at position (x, y). This will be a symbolic expression in terms of quantities defined on the figure.

24.5 Magnetic Fields Exert Forces on Moving Charges

17. For each of the following, draw the magnetic force vector on the charge or, if appropriate, write "$\vec{F}$ into page," "$\vec{F}$ out of page," or "$\vec{F} = \vec{0}$."

a.

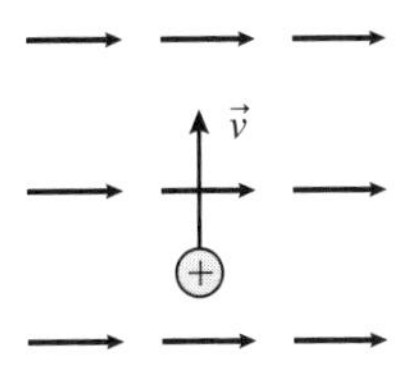

b.

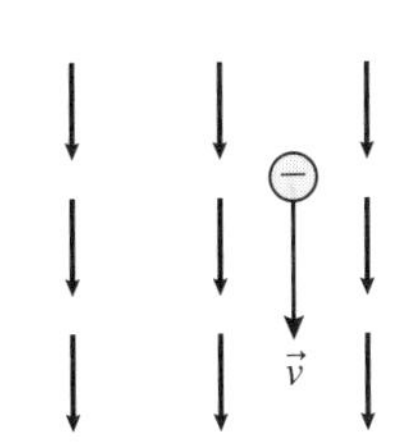

c.

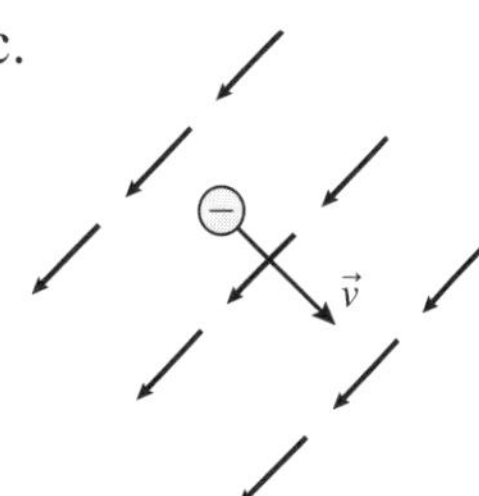

d.

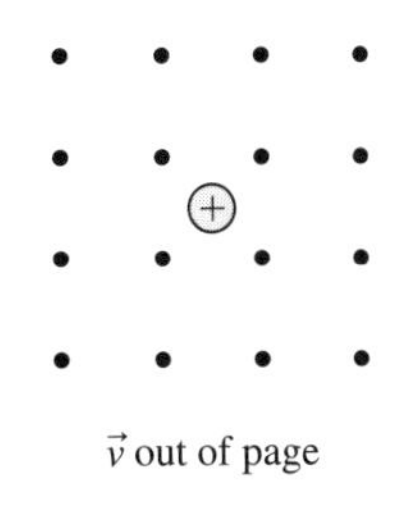

e.

f.

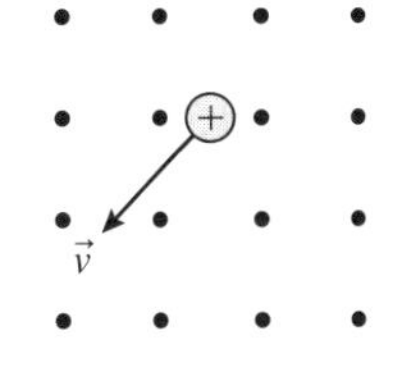

18. For each of the following, is the charge positive or negative? Write a + or a – on the charge.

a.

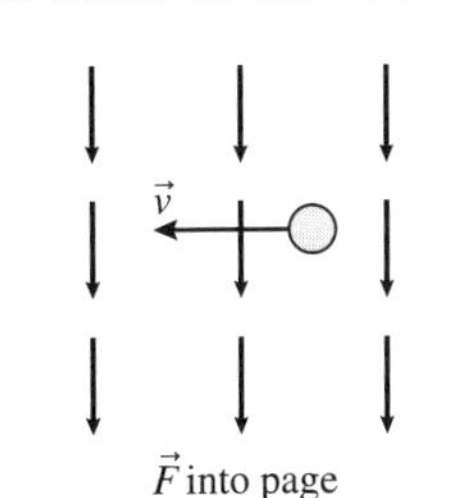

b.

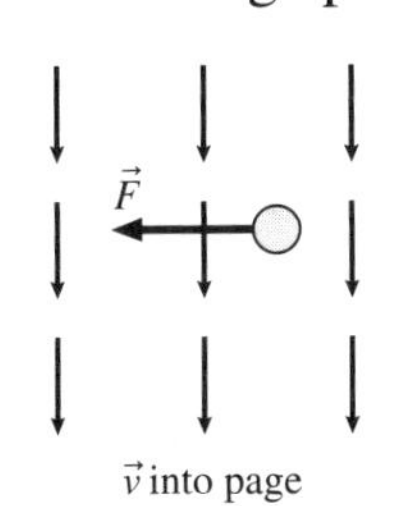

c.

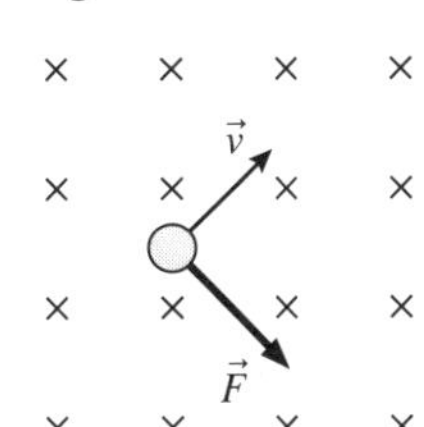

d.

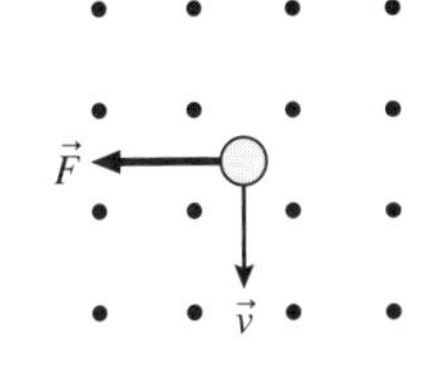

19. A positive ion, initially traveling into the page, is shot through the gap in a magnet. Is the ion deflected up, down, left, or right? Explain.

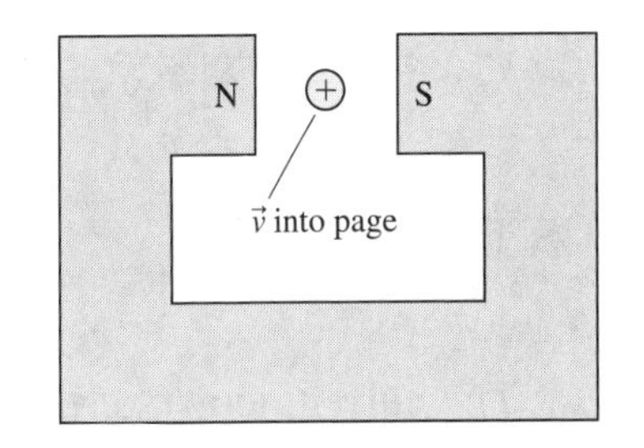

20. A positive ion is shot between the plates of a parallel-plate capacitor.

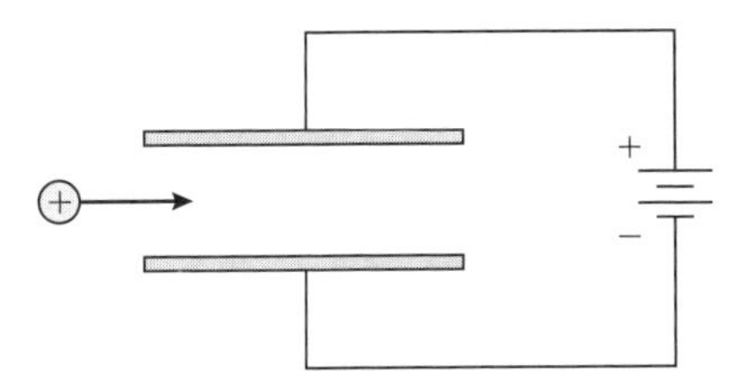

a. In what direction is the electric force on the ion?

b. Could a magnetic field exert a magnetic force on the ion that is opposite in direction to the electric force? If so, show the magnetic field on the figure. If not, why not?

24.6 Magnetic Fields Exert Forces on Currents

24.7 Magnetic Fields Exert Torques on Dipoles

21. Three current-carrying wires are perpendicular to the page. Construct a force vector diagram on the figure to find the net force on the upper wire due to the two lower wires.

22. Three current-carrying wires are perpendicular to the page.
    a. Construct a force vector diagram on each wire to determine the direction of the net force on each wire.
    b. Can three *charges* be placed in a triangular pattern so that their force diagram looks like this? If so, draw it. If not, why not?

23. A current-carrying wire passes between two bar magnets. Is there a net force on the wire? If so, draw the force vector. If not, why not?

a. S N ⊗ S N

b. S N ⊗ N S

24. The current loop exerts a repulsive force on the bar magnet. On the figure, show the direction of the current in the loop. Explain.

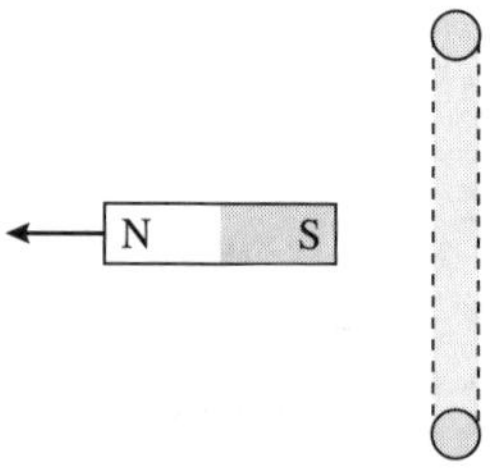

N S

25. The south pole of a bar magnet is held near a current loop. Does the bar magnet attract the loop, repel the loop, or have no effect on the loop? Explain.

26. A current loop is placed between two bar magnets. Does the loop move to the right, move to the left, rotate clockwise, rotate counterclockwise, some combination of these, or none of these? Explain.

27. A square current loop is placed in a magnetic field as shown.

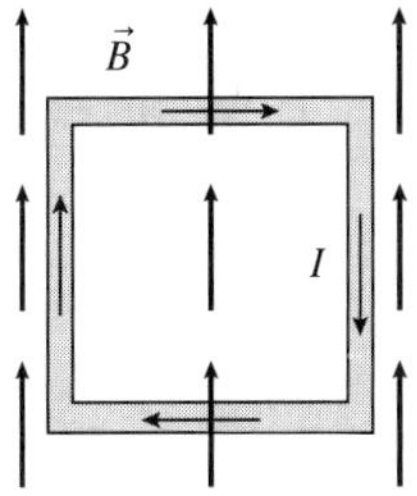

a. Does the loop undergo a displacement? If so, is it up, down, left, or right? If not, why not?

b. Does the loop rotate? If so, which edge rotates out of the page and which edge into the page? If not, why not?

28. A long vertical wire, attached by two springs to a vertical wall, passes through a region of uniform magnetic field of height L. We would like to know what current in the wire will cause the springs to be stretched an amount x. Assume that something unseen supports the weight of the wire.

PSA 24.1

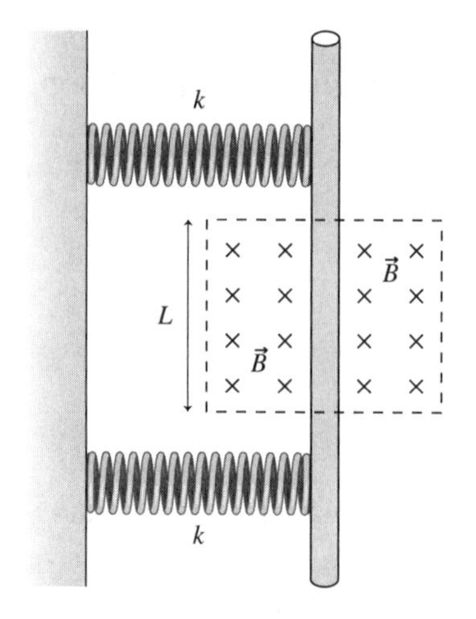

a. If the springs are stretched, what is the direction of the magnetic force on the wire? Draw it on the figure and label it $\vec{F}_{\text{wire}}$.

b. What direction is the current I in the wire? Draw and label it on the figure, and explain your choice.

c. Draw and label the forces of the springs on the wire. Call them each $\vec{F}_{\text{sp}}$. Make sure they have the proper lengths relative to the force you drew in part a.

d. If the wire is in equilibrium, what is the relationship between the forces you've drawn? This should be a mathematical statement.

e. Write expressions for any forces to the left. These will be symbolic expressions in terms of quantities defined on the figure and in the discussion above.

f. Now write symbolic expressions for any forces to the right.

g. Insert your expressions of parts e and f into the mathematical relationship of part d.

h. Complete your solution by solving this equation for I.

You Write the Problem!

Exercises 29–32: You are given the equation that is used to solve a problem. For each of these:

a. Write a *realistic* physics problem for which this is the correct equation. Look at worked examples and end-of-chapter problems in the textbook to see what realistic physics problems are like.
b. Finish the solution of the problem.

29. $1.5 \times 10^{-4}\ \text{T} = \dfrac{(1.257 \times 10^{-6}\ \text{T} \cdot \text{m/A})(15\ \text{A})}{2\pi r}$

30. $0.015\ \text{m} = \dfrac{(9.11 \times 10^{-31}\ \text{kg})(1.5 \times 10^{7}\ \text{m/s})}{(1.60 \times 10^{-19}\ \text{C})B}$

31. $(1.67 \times 10^{-27}\ \text{kg})a = (1.60 \times 10^{-19}\ \text{C})(7.4 \times 10^{5}\ \text{m/s}) \times \dfrac{(1.257 \times 10^{-6}\ \text{T} \cdot \text{m/A})(150)(2.5\ \text{A})}{0.15\ \text{m}}$

32. $m(9.8\ \text{m/s}^2) = (5.5\ \text{A})(0.075\ \text{m})(0.55\ \text{T})$

25 EM Induction and EM Waves

25.1 Induced Currents

25.2 Motional emf

1. The figures below show one or more metal wires sliding on fixed metal rails in a magnetic field. For each, determine if the induced current flows clockwise, flows counterclockwise, or is zero. Show your answer by drawing it.

a.

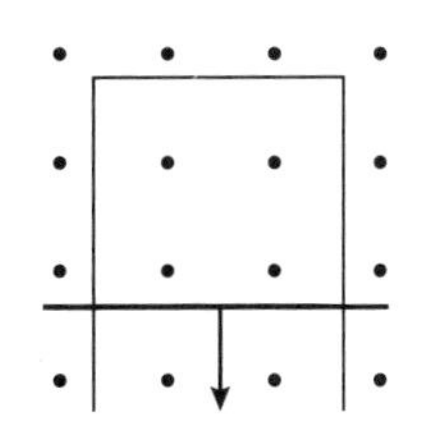

b.

c.

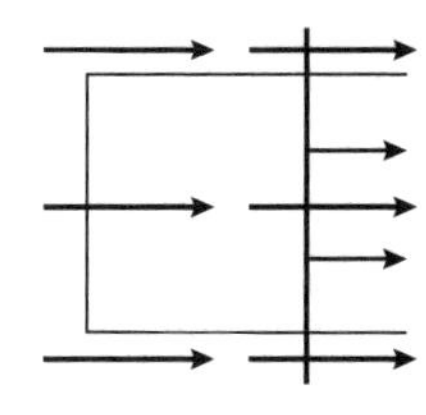

d.

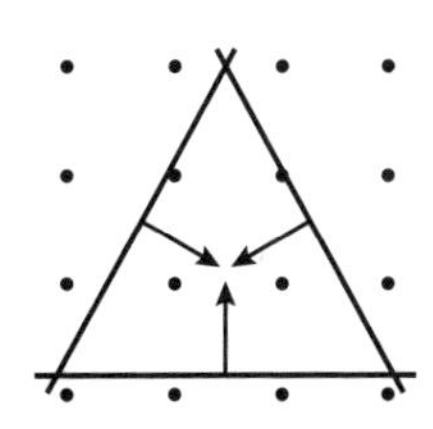

e.

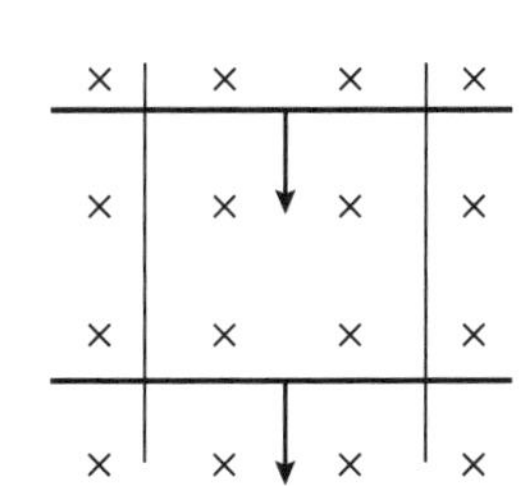

f.

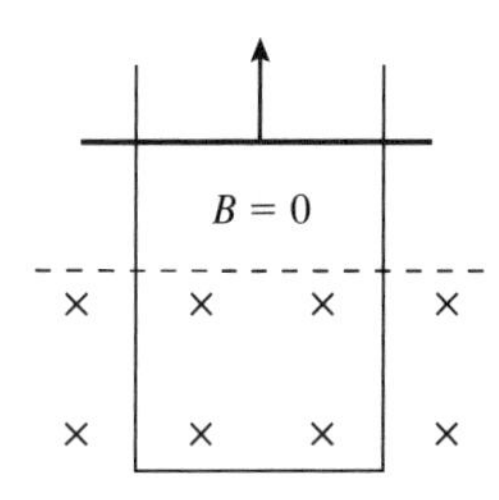

2. A loop of copper wire is being pulled from between two magnetic poles.

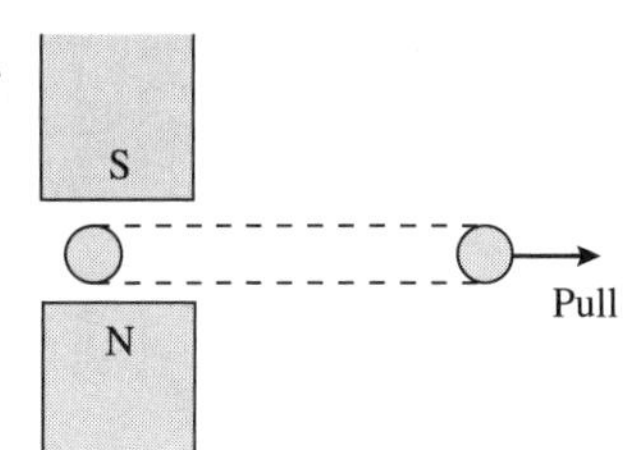

a. Show on the figure the current induced in the loop. Explain your reasoning.

b. Does either side of the loop experience a magnetic force? If so, draw a vector arrow or arrows on the figure to show any forces.

3. A vertical, rectangular loop of copper wire is half in and half out of a horizontal magnetic field (shaded gray). The field is zero beneath the dashed line. The loop is released and starts to fall.

   a. Add arrows to the figure to show the direction of the induced current in the loop.

   b. Is there a net magnetic force on the loop? If so, in which direction? Explain.

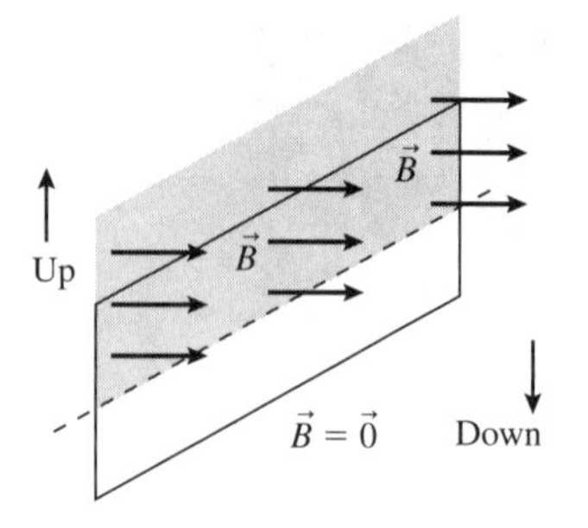

4. A metal bar rotates counterclockwise in a magnetic field as shown. Does the bar have a motional emf? If not, why not? If so, is the outer end of the bar positive or negative? Explain.

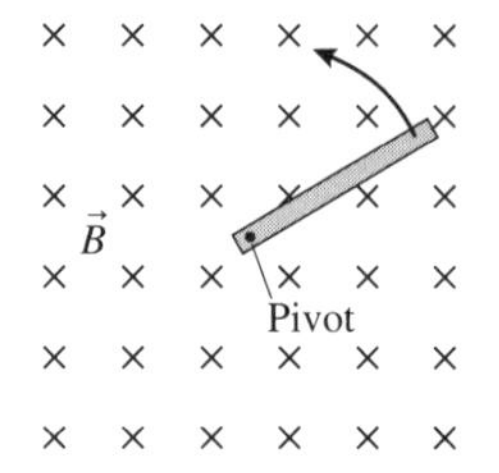

5. A metal bar oscillates back and forth on a spring as shown. Let the motional emf of the bar be positive when the top of the bar is positive and negative when the top of the bar is negative. Draw a graph showing approximately how the emf of the bar changes with time. Be sure your emf graph aligns correctly with the position graph above it.

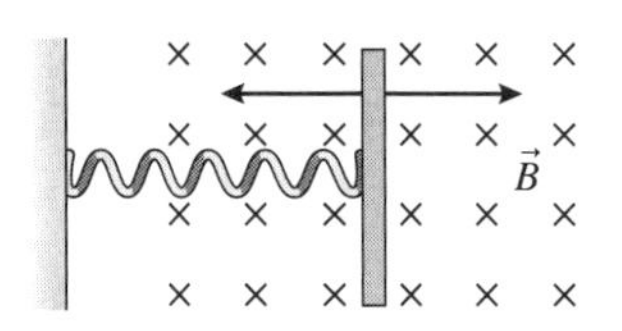

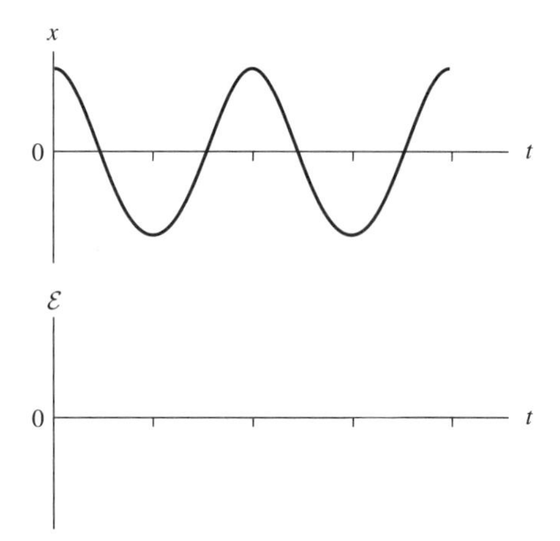

25.3 Magnetic Flux and Lenz's Law

25.4 Faraday's Law

6. The figure shows five loops in a magnetic field. The numbers indicate the lengths of the sides and the strength of the field. Rank in order, from largest to smallest, the magnetic fluxes Φ_1 to Φ_5. Some may be equal.

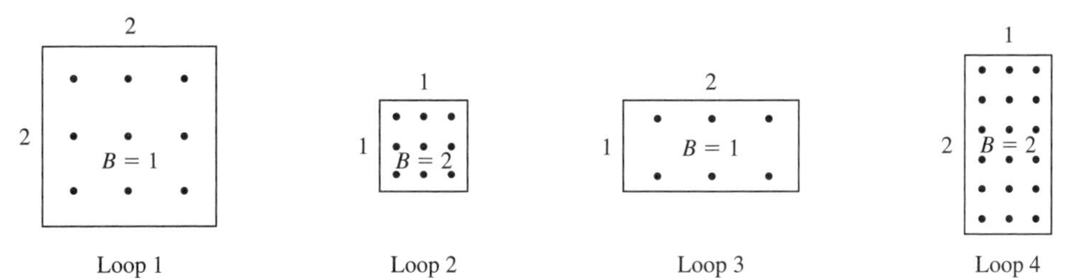

Order:

Explanation:

7. A circular loop rotates at constant speed about an axle through the center of the loop. The figure shows an edge view and defines the angle ϕ, which increases from 0° to 360° as the loop rotates.

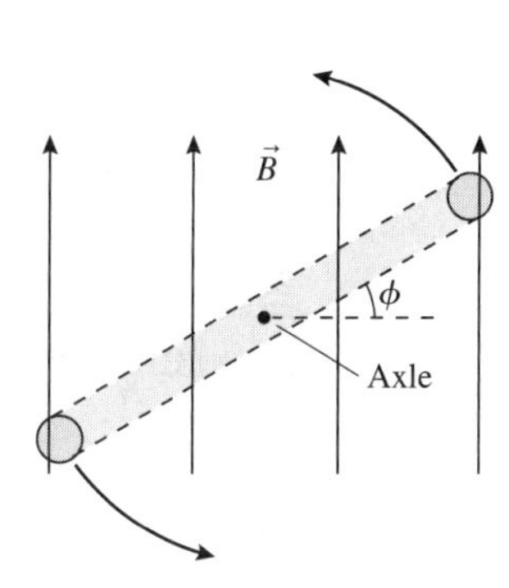

a. At what angle or angles is the magnetic flux a maximum?

b. At what angle or angles is the magnetic flux a minimum?

c. At what angle or angles is the magnetic flux *changing* most rapidly? Explain your choice.

8. A loop of wire is horizontal. A bar magnet is pushed toward the loop from below, along the axis of the loop.

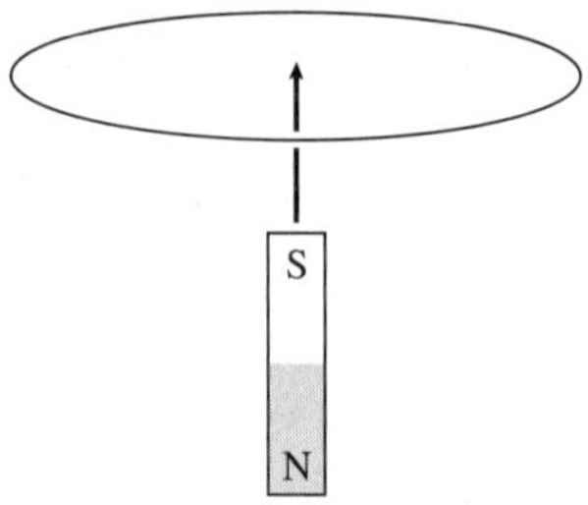

   a. What is the current direction in the loop as the magnet is approaching? Explain.

   b. Is there a magnetic force on the loop? If so, in which direction? Explain. Hint: A current loop is a magnetic dipole.

9. Does the loop of wire have a clockwise current, a counterclockwise current, or no current under the following circumstances? Explain.

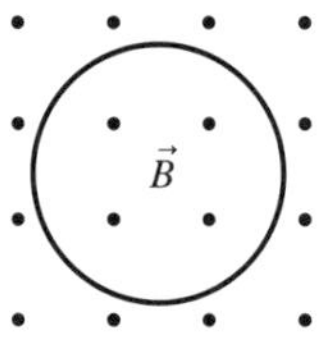

   a. The magnetic field points out of the page and its strength is increasing.

   b. The magnetic field points out of the page and its strength is constant.

   c. The magnetic field points out of the page and its strength is decreasing.

10. A loop of wire is perpendicular to a magnetic field. The magnetic field strength as a function of time is given by the top graph. Draw a graph of the current in the loop as a function of time. Let a positive current represent a current that comes out of the top of the loop and enters the bottom of the loop. There are no numbers for the vertical axis, but your graph should have the correct shape and proportions.

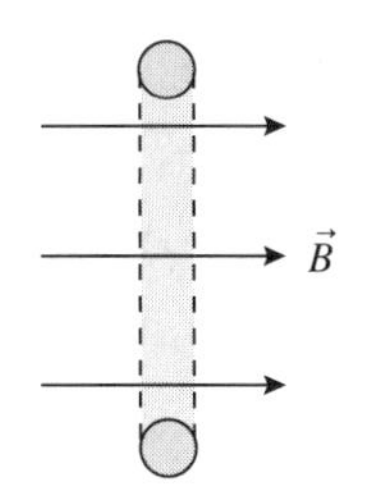

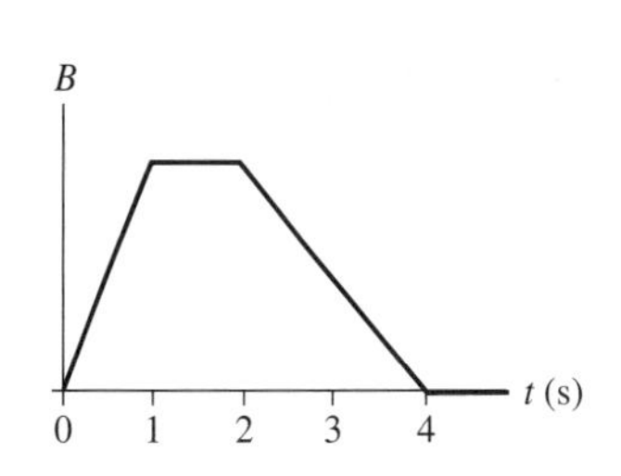

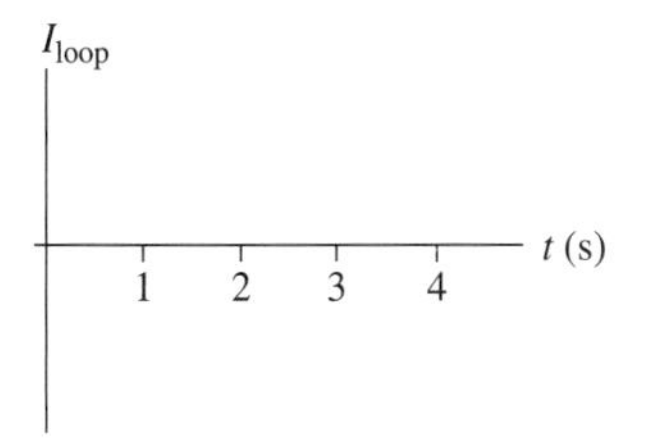

11. For the following questions, consider the current in the loop to be positive if it comes out of the top of the loop and enters the bottom; negative if it comes out of the bottom of the loop and enters the top.

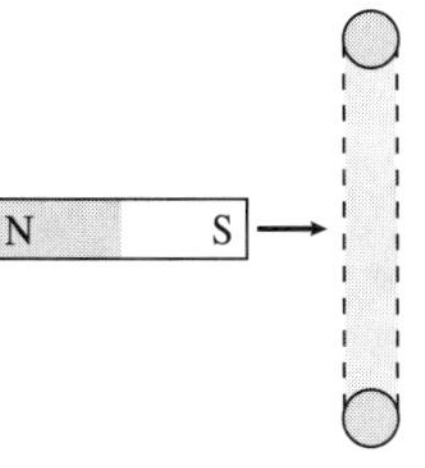

a. As the magnet is pushed into the loop, is the current in the loop positive, negative, or zero? Explain.

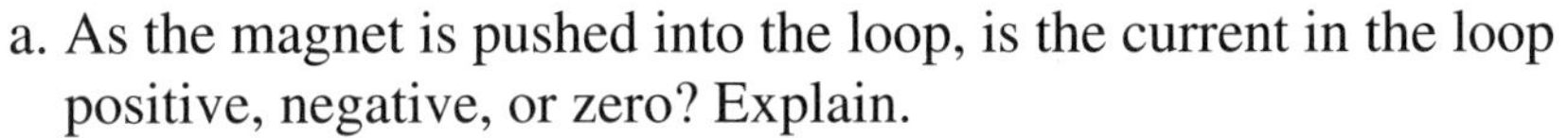

b. As the magnet is held at rest in the center of the loop, is the current in the loop positive, negative, or zero? Explain.

c. As the magnet is withdrawn from the loop, pulled back to the left, is the current in the loop positive, negative, or zero? Explain.

d. If the magnet is pushed into the loop more rapidly than in part a, does the size of the current increase, decrease, or stay the same? Explain.

You Write the Problem!

Exercises 12–14: You are given the equation that is used to solve a problem. For each of these:

a. Write a *realistic* physics problem for which this is the correct equation. Look at worked examples and end-of-chapter problems in the textbook to see what realistic physics problems are like.
b. Finish the solution of the problem.

12. $2.6\text{ mV} = v(15\text{ cm})(5.0 \times 10^{-5}\text{ T})$

13. $\pi(0.040\text{ m})^2(0.12\text{ T})\cos\theta = 5.2 \times 10^{-4}\text{ Wb}$

14. $0.25\text{ A} = \dfrac{(0.15\text{ m})^2(0.30\text{ T})/(0.50\text{ s})}{R}$

15. The graph shows how the magnetic field changes through a rectangular loop of wire with resistance R. Draw a graph of the current in the loop as a function of time. Let a counterclockwise current be positive, a clockwise current be negative.

PSA 25.1

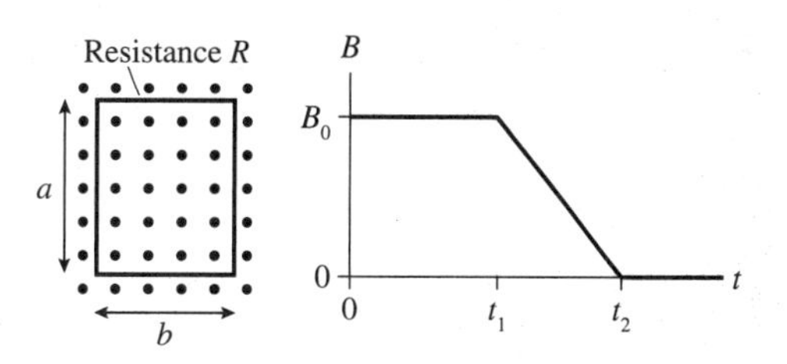

a. What is the magnetic flux through the loop at $t = 0$? ____

b. Does this flux *change* between $t = 0$ and $t = t_1$? ____

c. Is there an induced current in the loop between $t = 0$ and $t = t_1$? ____

d. What is the magnetic flux through the loop at $t = t_2$? ____

e. What is the *change* in flux through the loop between t_1 and t_2? ____

f. What is the time interval between t_1 and t_2? ____

g. What is the magnitude of the induced emf between t_1 and t_2? ____

h. What is the magnitude of the induced current between t_1 and t_2? ____

i. Does the magnetic field point out of or into the loop? ____

f. Between t_1 and t_2, is the magnetic flux increasing or decreasing? ____

g. To oppose the *change* in the flux between t_1 and t_2, should the magnetic field of the induced current point out of or into the loop? ____

h. Is the induced current between t_1 and t_2 positive or negative? ____

i. Does the flux through the loop change after t_2? ____

j. Is there an induced current in the loop after t_2? ____

k. Use all this information to draw a graph of the induced current. Add appropriate labels on the vertical axis.

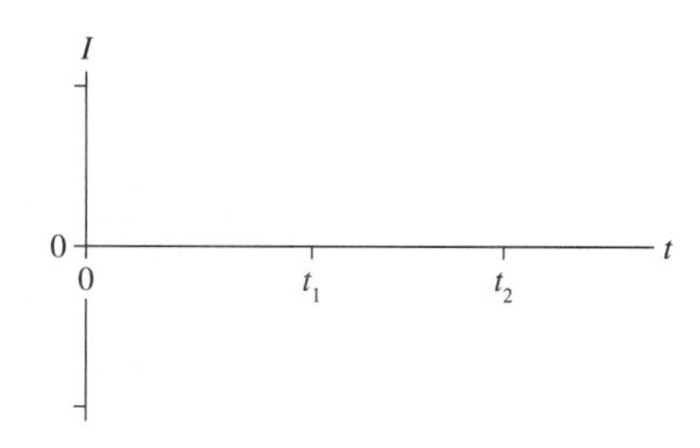

25.5 Electromagnetic Waves

16. The intensity of an electromagnetic wave is 10 W/m^2. What will be the intensity if:
    a. The amplitude of the electric field is doubled?

    b. The amplitude of the magnetic field is doubled?

    c. The frequency is doubled?

17. The intensity of a polarized electromagnetic wave is 10 W/m^2. What will be the intensity of the wave after it passes through a polarizing filter whose axis makes the following angle with the plane of polarization?

    $\theta = 0°$ ____________ $\theta = 60°$ ____________

    $\theta = 30°$ ____________ $\theta = 90°$ ____________

    $\theta = 45°$ ____________

18. A polarized electromagnetic wave passes through a polarizing filter. Draw the electric field of the wave after it has passed through the filter.

    a. b.

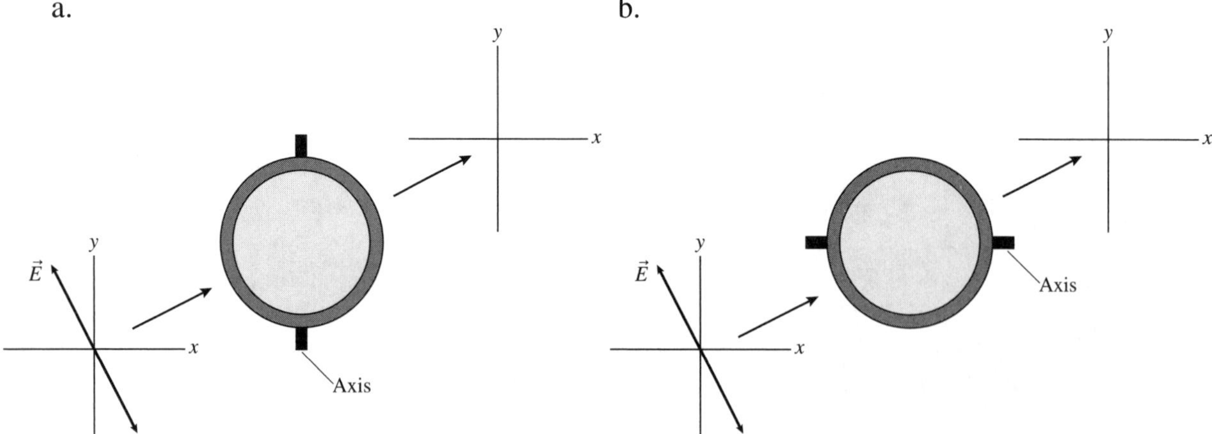

19. A polarized electromagnetic wave passes through a series of polarizing filters. Draw the electric field of the wave after it has passed through each filter or, if appropriate, state that $E = 0$.

a.

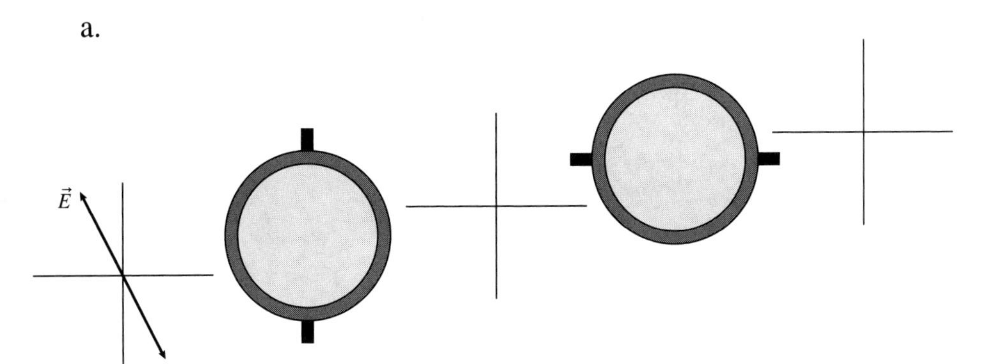

b.

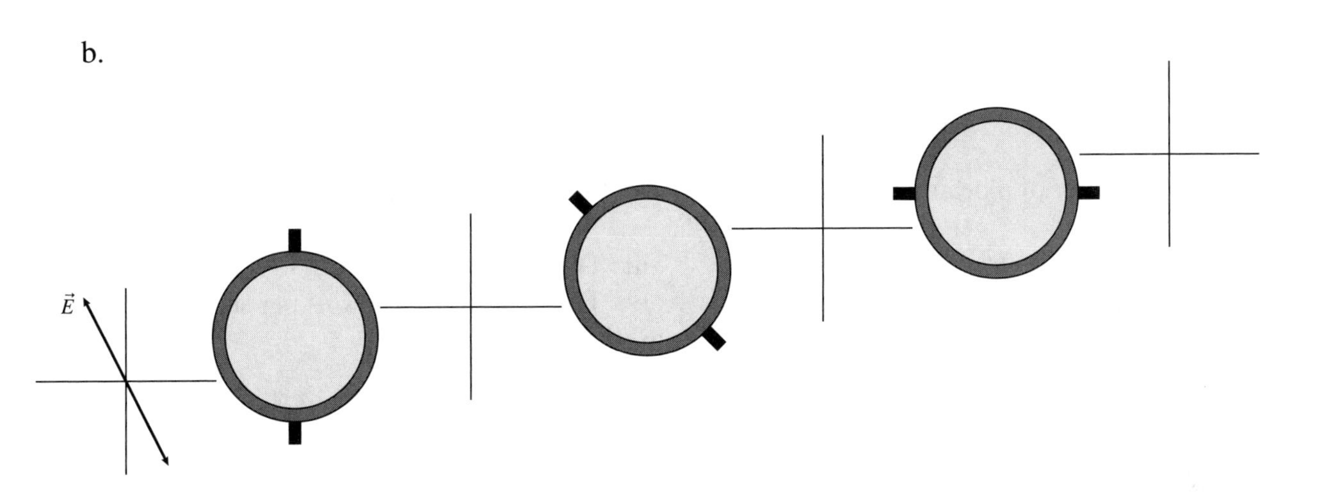

25.6 The Photon Model of Electromagnetic Waves

20. Three laser beams have wavelengths $\lambda_1 = 400$ nm, $\lambda_2 = 600$ nm, and $\lambda_3 = 800$ nm. The power of each laser beam is 1 W.

   a. Rank in order, from largest to smallest, the energies E_1, E_2, and E_3 of the photons in these three laser beams.

   Order:

   Explanation:

   b. Rank in order, from largest to smallest, the number of photons per second N_1, N_2, and N_3 delivered by the three laser beams.

   Order:

   Explanation:

21. The intensity of a beam of light is increased, but the light's frequency is unchanged. As a result:

   i. The photons travel faster.
   ii. Each photon has more energy.
   iii. The photons are larger.
   iv. There are more photons per second.

   Which of these (perhaps more than one) are true? Explain.

22. The frequency of a beam of light is increased, but the light's intensity is unchanged. As a result:

   i. The photons travel faster.
   ii. Each photon has more energy.
   iii. There are fewer photons per second.
   iv. There are more photons per second.

   Which of these (perhaps more than one) are true? Explain.

23. Light of wavelength $\lambda = 1\ \mu$m is emitted from point A. A photon is detected 5 μm away at point B. On the figure, draw the trajectory that a photon follows between points A and B.

A

B

5 μm

25.7 The Electromagnetic Spectrum

24. The graph at the right shows the thermal emission spectrum of light as a function of wavelength from an object at temperature T. On the graph, sketch approximately how the spectrum would appear if the object's absolute temperature were doubled to $2T$.

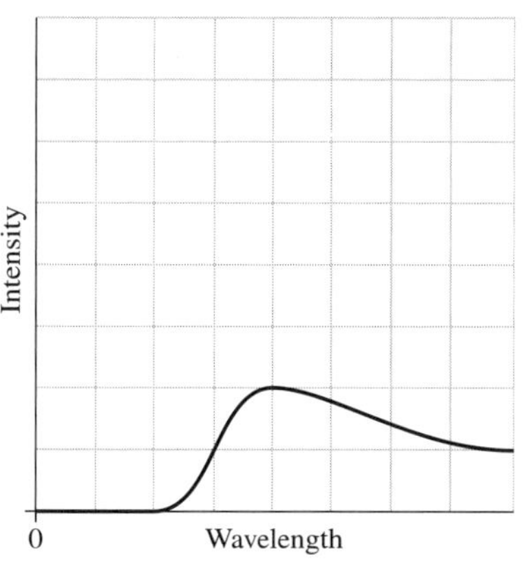

25. An object's thermal emission spectrum has peak intensity at a wavelength of 3.0 μm. What will be the wavelength of peak intensity if the object's absolute temperature is tripled?

26. An astronomer observes the thermal emission spectrum of two stars. Star A has maximum intensity at a wavelength of 400 nm. Star B has maximum intensity at 1200 nm. What is the ratio T_A/T_B of the absolute temperatures of the two stars?

26 AC Electricity

26.1 Alternating Current

1. The graph shows two cycles of the AC current in a simple AC resistor circuit.

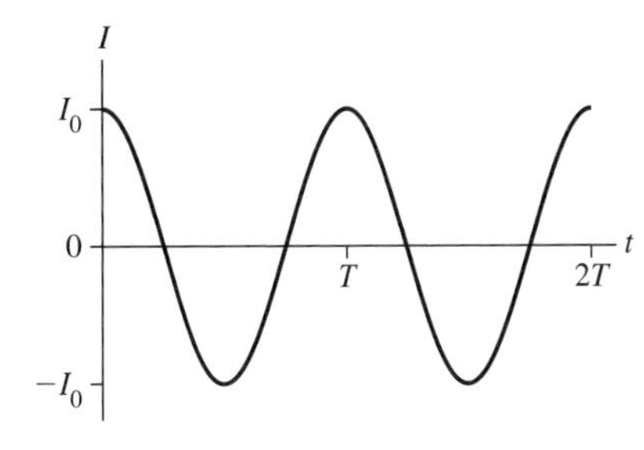

   a. What is the *average* value of the current after two whole cycles?
   b. On the axes below the current graph, draw a graph of the *square* of the current I^2. Make sure your graphs features are aligned with those of the AC current graph above it.

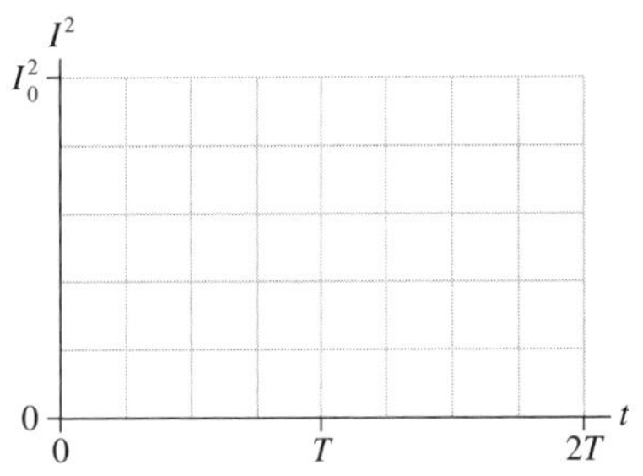

   c. If the frequency of the AC current is f, what is the frequency of the square of the current?
   d. Based on your graph, what is the *average* value of the square of the current after two whole cycles?
   e. What is the square root of your answer to part d? That is, the square root of the average of the square of the current? This value is the root-mean-square current I_{rms}.

2. The peak current through the resistor in an AC circuit is 4.0 A. What is the peak current if:
   a. The resistance R is doubled?
   b. The peak voltage V_R is doubled?
   c. The frequency f is doubled?

26.2 AC Electricity and Transformers

3. Your phone's 12 V battery is dead. It's a holiday and all the stores are closed. You do have two 1.5 V batteries from your camera and a transformer with $N_1 = 100$ turns on the primary and $N_2 = 400$ turns on the secondary. Can you use the batteries and transformer to get your phone operating? If so, how would you do it? If not, why not?

4. The primary coil of a transformer draws a 4.0 A rms current when plugged into a 120 V outlet. The secondary voltage is 40 V. What current does the secondary coil of the transformer deliver to the load?

26.3 Household Electricity

26.4 Biological Effects and Electrical Safety

5. The figure shows a standard North American household electric socket.

a. Label the socket holes. One is "hot," the other "neutral."
b. When you plug a device into the socket, does the current always come out of one hole and go back into the other? If so, which hole does the current come out of? If not, why not?

26.5 Capacitor Circuits

6. Current and voltage graphs are shown for a capacitor circuit with $f = 1000$ Hz.

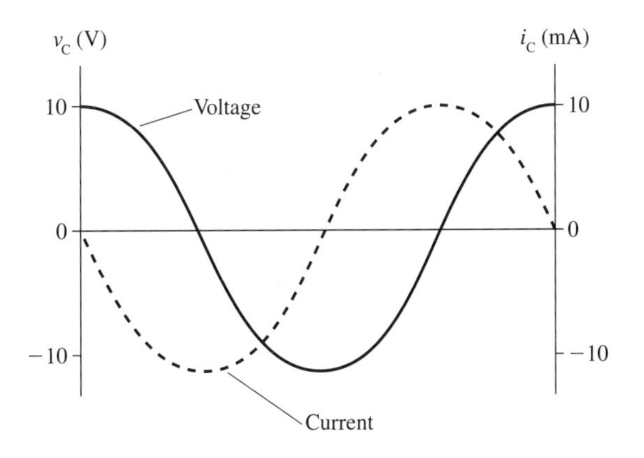

   a. What is the capacitive reactance X_C?

   b. What is the capacitance C?

7. Consider these three circuits.

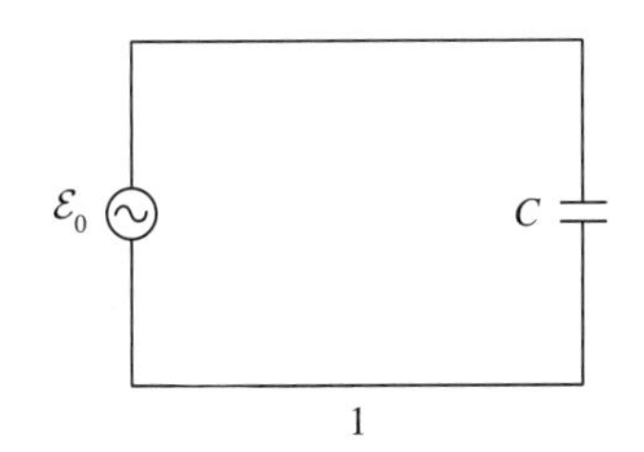

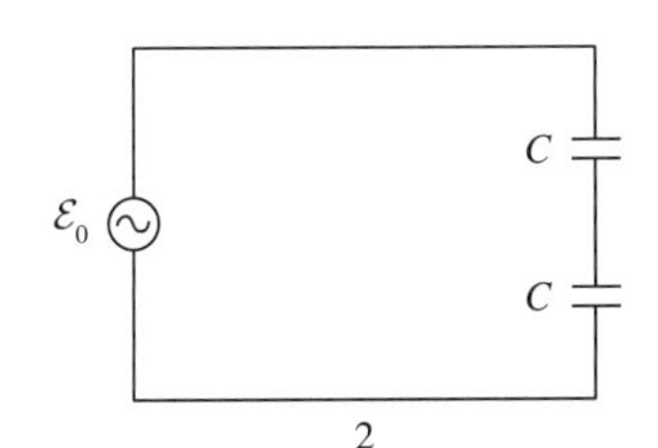

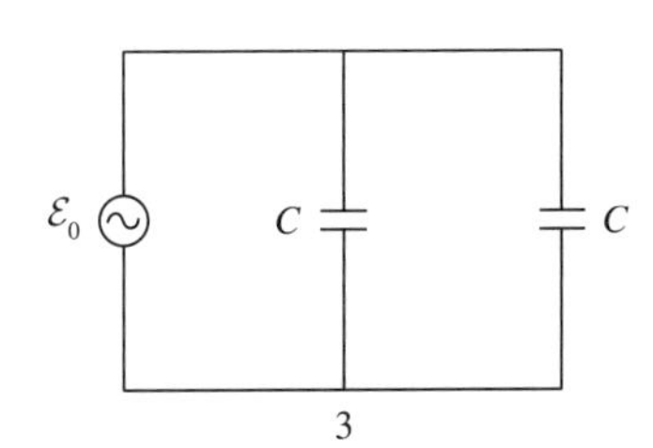

Rank in order, from largest to smallest, the peak currents $(I_C)_1$ to $(I_C)_3$.

Order:

Explanation:

8. The peak current through the capacitor in an AC circuit is 4.0 A. What is the peak current if:
   a. The capacitance C is doubled?

   b. The peak voltage V_C is doubled?

   c. The frequency f is doubled?

9. A 13 μF capacitor is connected to a 5.0 V/250 Hz oscillating voltage. What is the instantaneous capacitor current when the instantaneous capacitor voltage is $v_C = -5.0$ V? Explain. Note that no calculations are required, only careful reasoning.

26.6 Inductors and Inductor Circuits

10. Current and voltage graphs are shown for an inductor circuit with $f = 1000$ Hz.

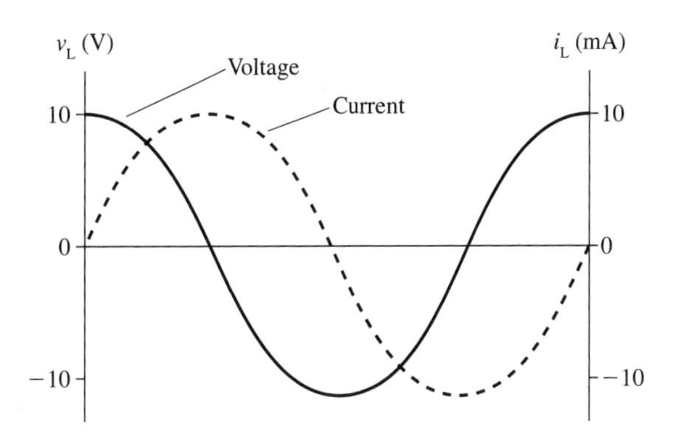

a. What is the inductive reactance X_L?

b. What is the inductance L?

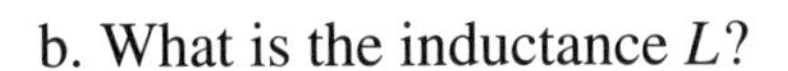

11. The peak current through the inductor in an AC circuit is 4.0 A. What is the peak current if:

a. The inductance L is doubled?

b. The peak voltage V_L is doubled?

c. The frequency f is doubled?

12. Consider these three circuits.

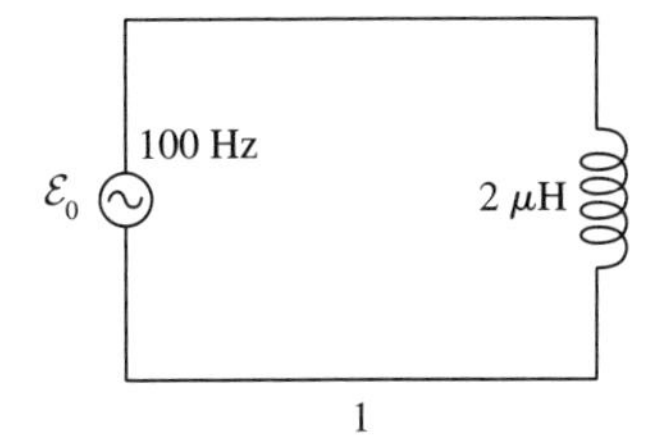

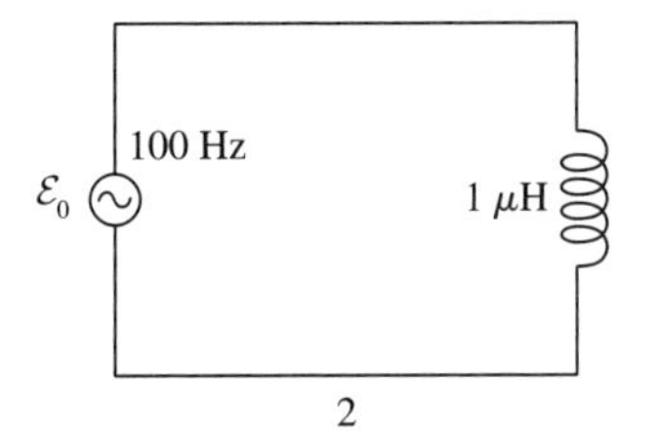

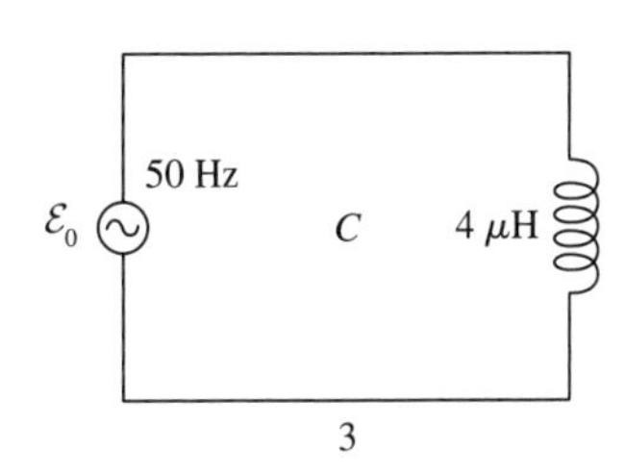

Rank in order, from largest to smallest, the peak currents $(I_L)_1$ to $(I_L)_3$.

Order:

Explanation:

26.7 Oscillation Circuits

13. The resonance frequency of an *RLC* circuit is 1000 Hz. What is the resonance frequency if:
   a. The resistance R is doubled?

   b. The inductance L is doubled?

   c. The capacitance C is doubled?

   d. The peak emf $\mathcal{E}_0$ is doubled?

14. Two *RLC* circuits have the same resistance and same capacitance but different inductances. The graph shows the current oscillation at the resonance frequency of each circuit. Is L_1 for circuit 1 larger or smaller than L_2 for circuit 2? Explain.

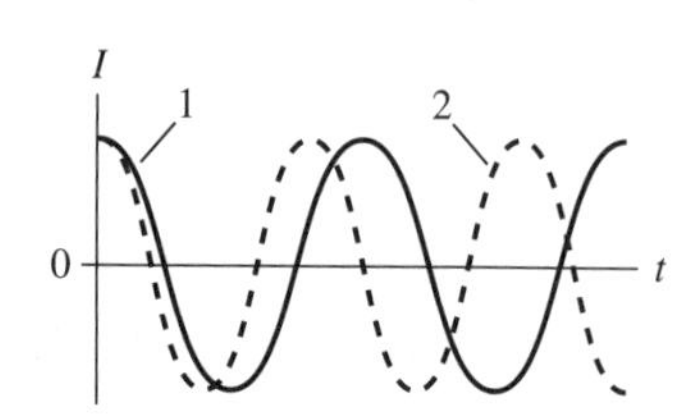

27 Relativity

27.1 Relativity: What's It All About?

27.2 Galilean Relativity

1. In which reference frame, S or S′, does the ball move faster?

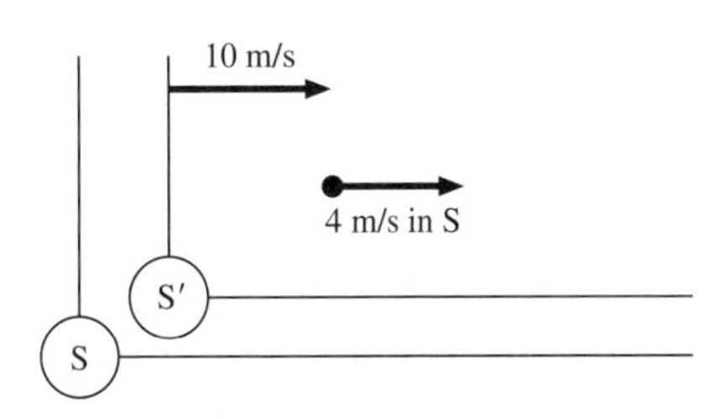

2. Frame S′ moves parallel to the x-axis of frame S.

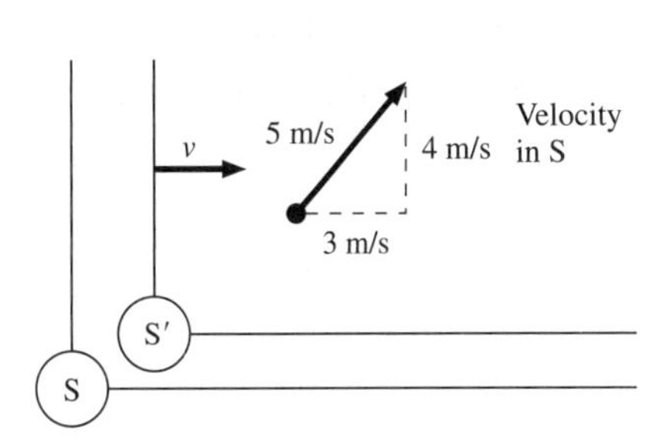

a. Is there a value of v for which the ball is at rest in S′? If so, what is v? If not, why not?

b. Is there a value of v for which the ball has a minimum speed in S′? If so, what is v? If not, why not?

3. What are the speed and direction of each ball in a reference frame that moves to the right at 2 m/s?

4. Anita is running to the right at 5 m/s. Balls 1 and 2 are thrown toward her at 10 m/s by friends standing on the ground. According to Anita, which ball is moving faster? Or are both speeds the same? Explain.

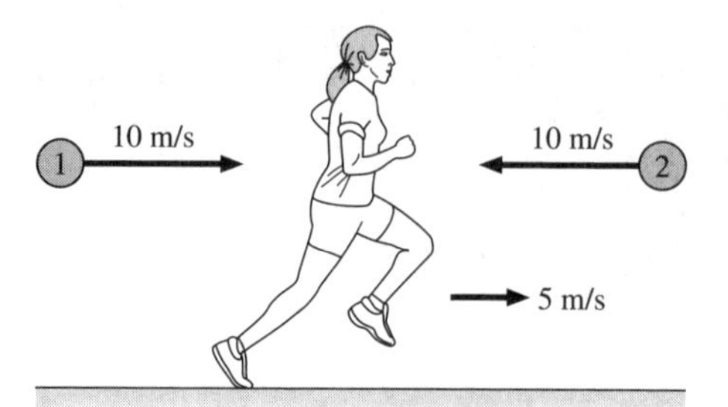

5. Anita is running to the right at 5 m/s. Balls 1 and 2 are thrown toward her by friends standing on the ground. According to Anita, both balls are approaching her at 10 m/s. Which ball was thrown at a faster speed? Or were they thrown with the same speed? Explain.

27.3 Einstein's Principle of Relativity

6. Teenagers Sam and Tom are playing chicken in their rockets. As seen from the earth, each is traveling at $0.95c$ as he approaches the other. Sam fires a laser beam toward Tom.

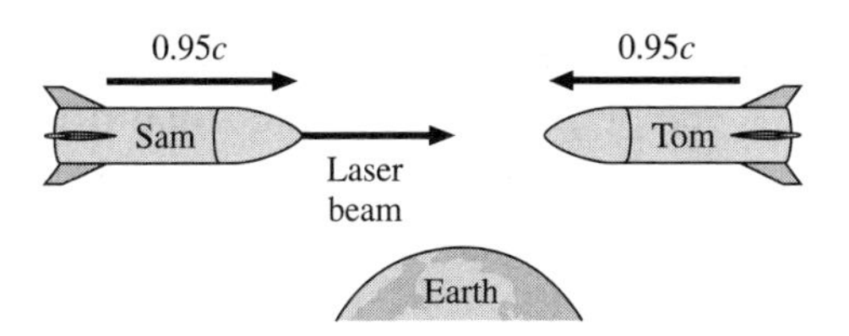

   a. What is the speed of the laser beam relative to Sam?

   b. What is the speed of the laser beam relative to Tom?

27.4 Events and Measurements

7. It is a bitter cold day at the South Pole, so cold that the speed of sound is only 300 m/s. The speed of light, as always, is 300 m/μs. A firecracker explodes 600 m away from you.

   a. How long after the explosion until you see the flash of light? ________

   b. How long after the explosion until you hear the sound? ________

   c. Suppose you see the flash at $t = 2.000002$ s. At what time was the explosion? ________

   d. What are the spacetime coordinates for the event "firecracker explodes"? Assume that you are at the origin and that the explosion takes place at a position on the positive x-axis.

8. You are at the origin of a coordinate system containing clocks, but you're not sure if the clocks have been synchronized. The clocks have reflective faces, allowing you to read them by shining light on them. You flash a bright light at the origin at the instant your clock reads $t = 2.000000$ s.

   a. At what time will you see the reflection of the light from a clock at $x = 3000$ m?

   b. When you see the clock at $x = 3000$ m, it reads 2.000020 s. Is the clock synchronized with your clock at the origin? Explain.

27.5 The Relativity of Simultaneity

9. Firecracker 1 is 300 m from you. Firecracker 2 is 600 m from you in the same direction. You see both explode at the same time. Define event 1 to be "firecracker 1 explodes" and event 2 to be "firecracker 2 explodes." Does event 1 occur before, after, or at the same time as event 2? Explain.

10. Firecrackers 1 and 2 are 600 m apart. You are standing exactly halfway between them. Your lab partner is 300 m on the other side of firecracker 1. You see two flashes of light, from the two explosions, at exactly the same instant of time. Define event 1 to be "firecracker 1 explodes" and event 2 to be "firecracker 2 explodes." According to your lab partner, based on measurements he or she makes, does event 1 occur before, after, or at the same time as event 2? Explain.

11. Can two spatially separated events be simultaneous if they are seen at two different times? If not, why not? If so, give an example.

12. Can two simultaneous events, A and B, at different locations be seen by different people as taking place in a different order, so that one person sees A then B, but another person sees B followed by A? If not, why not? If so, give an example.

13. Two trees are 600 m apart. You are standing exactly halfway between them and your lab partner is at the base of tree 1. Lightning strikes both trees.
    a. Your lab partner, based on measurements he or she makes, determines that the two lightning strikes were simultaneous. What did you see? Did you see the lightning hit tree 1 first, hit tree 2 first, or hit them both at the same instant of time? Explain.

    b. Lightning strikes again. This time your lab partner sees both flashes of light at the same instant of time. What did you see? Did you see the lightning hit tree 1 first, hit tree 2 first, or hit them both at the same instant of time? Explain.

    c. In the scenario of part b, were the lightning strikes simultaneous? Explain.

14. A rocket is traveling from left to right. At the instant it is halfway between two trees, lightning simultaneously (in the rocket's frame) hits both trees.

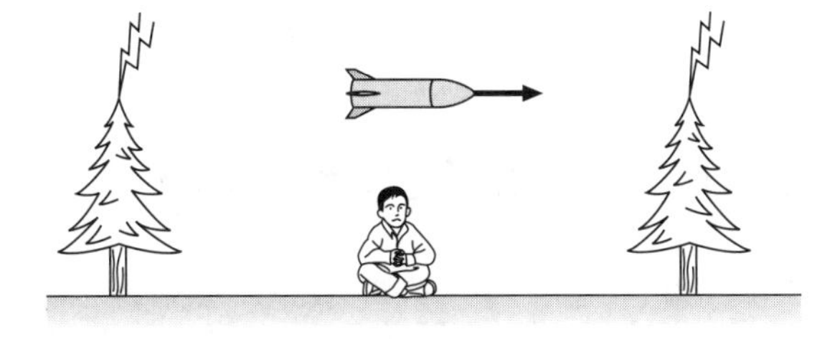

    a. Do the light flashes reach the rocket pilot simultaneously? If not, which reaches him first? Explain.

    b. A student was sitting on the ground halfway between the trees as the rocket passed overhead. According to the student, were the lightning strikes simultaneous? If not, which tree was hit first? Explain.

27.6 Time Dilation

15. Clocks C_1 and C_2 in frame S are synchronized. Clock C′ moves at speed v relative to frame S. Clocks C′ and C_1 read exactly the same as C′ goes past. As C′ passes C_2, is the time shown on C′ earlier than, later than, or the same as the time shown on C_2? Explain.

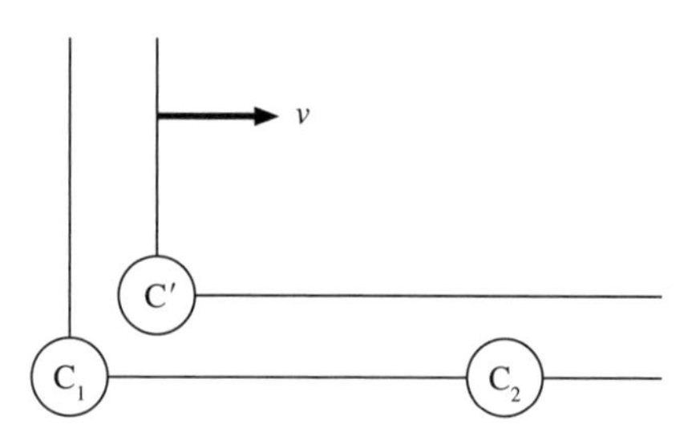

16. Your friend flies from Los Angeles to New York. She carries an accurate stopwatch with her to measure the flight time. You and your assistants on the ground also measure the flight time.

    a. Identify the two events associated with this measurement.

    b. Who, if anyone, measures the proper time? ______

    c. Who, if anyone, measures the shorter flight time? ______

27.7 Length Contraction

17. Your friend flies from Los Angeles to New York. He determines the distance using the tried-and-true $d = vt$. You and your assistants on the ground also measure the distance, using meter sticks and surveying equipment.

    a. Who, if anyone, measures the proper length? ______

    b. Who, if anyone, measures the shorter distance? ______

18. Experimenters in B's reference frame measure $L_A = L_B$. Do experimenters in A's reference frame agree that A and B have the same length? If not, which do they find to be longer? Explain.

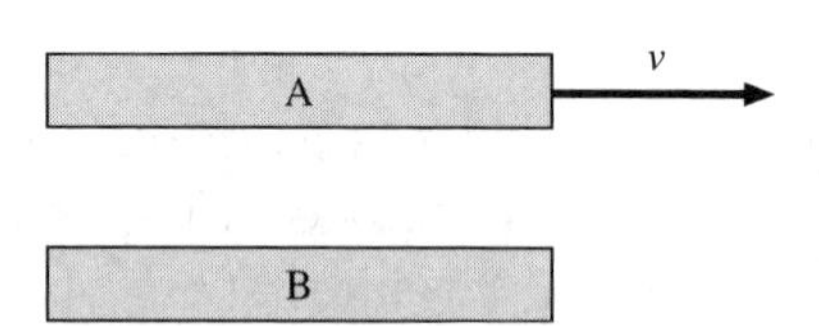

27.8 Velocities of Objects in Special Relativity

19. A rocket travels at speed $0.5c$ relative to the earth.

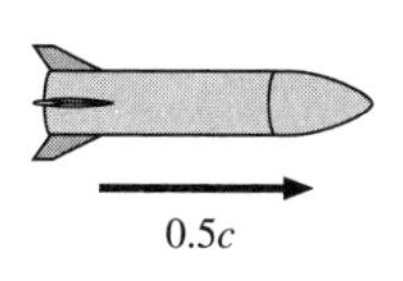

a. The rocket shoots a bullet in the forward direction at speed $0.5c$ relative to the rocket. Is the bullet's speed relative to the earth less than, greater than, or equal to c?

b. The rocket shoots a second bullet in the backward direction at speed $0.5c$ relative to the rocket. In the earth's frame, is the bullet moving right, moving left, or at rest?

27.9 Relativistic Momentum

20. Particle A has half the mass and twice the speed of particle B. Is p_A less than, greater than, or equal to p_B? Explain.

21. Particle A has one-third the mass of particle B. The two particles have equal momenta. Is u_A less than, greater than, or equal to $3u_B$? Explain.

22. Event B occurs at $t_B = 10.0\ \mu s$. An earlier event A, at $t_A = 5.0\ \mu s$, is the cause of B. What is the maximum possible distance that A can be from B?

27.10 Relativistic Energy

23. Can a particle of mass m have total energy less than mc^2? Explain.

24. Consider these 4 particles:

Particle	Rest energy	Total energy
1	A	A
2	B	$2B$
3	$2C$	$4C$
4	$3D$	$5D$

Rank in order, from largest to smallest, the particles' speeds u_1 to u_4.

Order:

Explanation:

28 Quantum Physics

28.1 X Rays and X-Ray Diffraction

1. Use trigonometry to show on the diagram that the path-length difference between the two x rays is $\Delta r = 2d\cos\theta$.

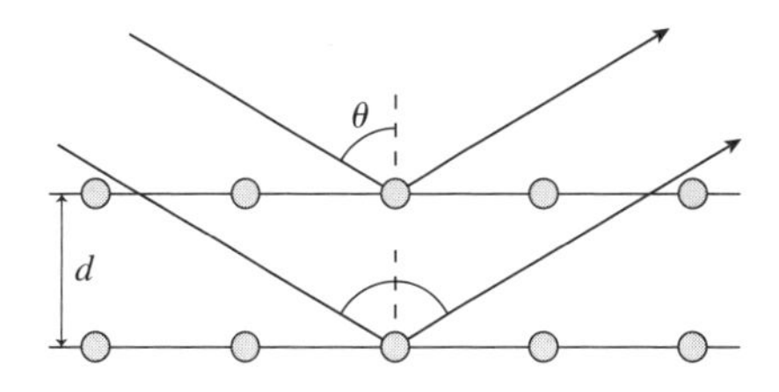

28.2 The Photoelectric Effect

2. a. A negatively charged electroscope can be discharged by shining an ultraviolet light on it. How does this happen?

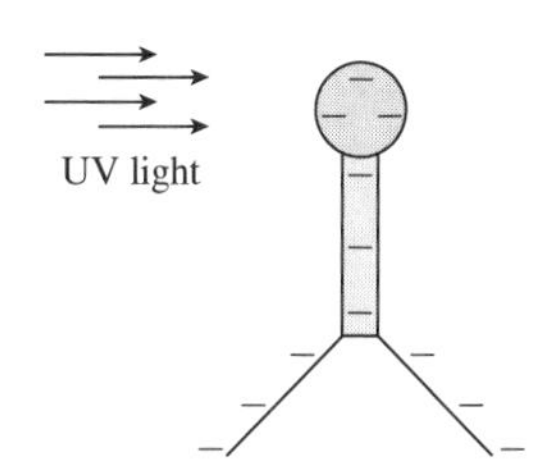

b. You might think that an ultraviolet light shining on an initially uncharged electroscope would cause the electroscope to become positively charged as photoelectrons are emitted. In fact, ultraviolet light has no noticeable effect on an uncharged electroscope. Why not?

3. In the photoelectric effect experiment, a current is measured while light is shining on the cathode. But this does not appear to be a complete circuit, so how can there be a current? Explain.

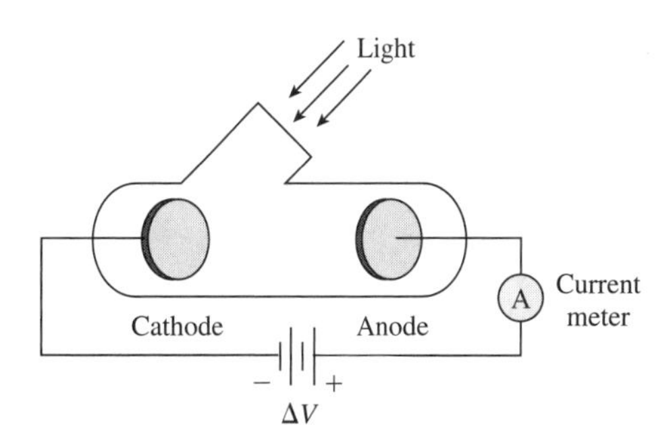

4. The work function of a metal measures:
   i. The kinetic energy of the electrons in the metal.
   ii. How tightly the electrons are bound within the metal.
   iii. The amount of work done by the metal when it expands.

   Which of these (perhaps more than one) are correct? Explain.

5. a. What is the significance of V_{stop}? That is, what have you learned if you measure V_{stop}? **Note:** Don't say that "$-V_{stop}$ is the potential that causes the current to stop." That is merely the definition of V_{stop}. It doesn't say what the *significance* of V_{stop} is.

   b. Why is it surprising that V_{stop} is independent of the light intensity? What would you *expect* V_{stop} to do as the intensity increases? Explain.

   c. If the wavelength of the light in a photoelectric effect experiment is increased, does V_{stop} increase, decrease, or stay the same? Explain.

6. The figure shows a typical current-versus-potential difference graph for a photoelectric effect experiment. On the figure, draw and label graphs for the following three situations:
   i. The light intensity is increased.
   ii. The light frequency is increased.
   iii. The cathode work function is increased.

   In each case, no other parameters of the experiment are changed.

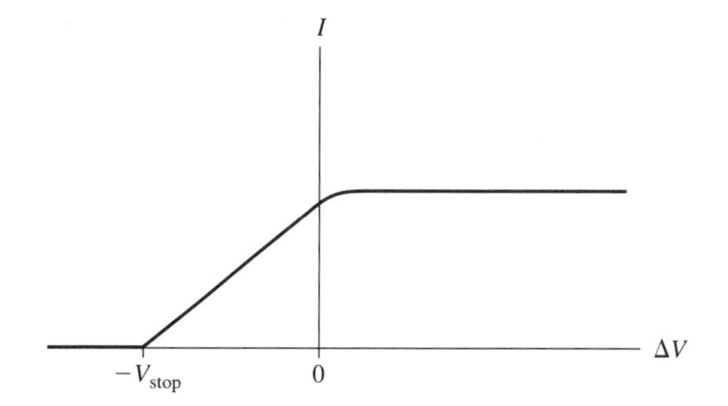

7. The figure shows a typical current-versus-frequency graph for a photoelectric effect experiment. On the figure, draw and label graphs for the following three situations:
   i. The light intensity is increased.
   ii. The anode-cathode potential difference is increased.
   iii. The cathode work function is increased.

   In each case, no other parameters of the experiment are changed.

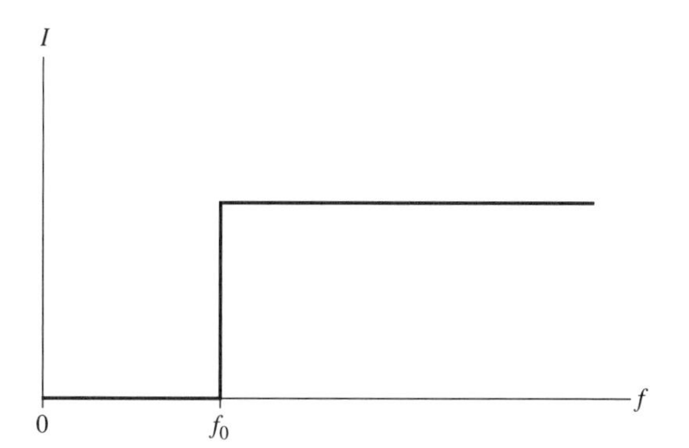

8. The figure shows a typical stopping potential-versus-frequency graph for a photoelectric effect experiment. On the figure, draw and label graphs for the following two situations:
   i. The light intensity is increased.
   ii. The cathode work function is increased.

   In each case, no other parameters of the experiment are changed.

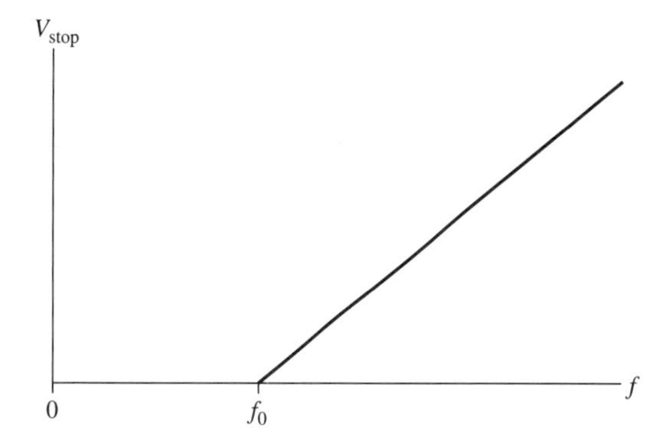

9. In a photoelectric effect experiment, the wavelength of the light is increased while the intensity of the light is held constant. As a result:

   i. There are more photoelectrons.

   ii. The photoelectrons are faster.

   iii. Both i and ii.

   iv. Neither i nor ii.

   Explain your choice.

10. In a photoelectric effect experiment, the intensity of light is increased while the wavelength of light is held constant. As a result:

    i. There are more photoelectrons.

    ii. The photoelectrons are faster.

    iii. Both i and ii.

    iv. Neither i nor ii.

    Explain your choice.

11. A gold cathode (work function = 5.1 eV) is illuminated with light of wavelength 250 nm. It is found that the photoelectron current is zero when $\Delta V = 0$ V. Would the current change if:

    a. The intensity is doubled?

    b. The anode-cathode potential difference is increased to $\Delta V = 5.5$ V?

    c. The cathode is changed to aluminum (work function = 4.3 eV)?

28.3 Photons

12. The top figure is the *negative* of the photograph of a single-slit diffraction pattern. That is, the darkest areas in the figure were the brightest areas on the screen. This photo was made with an extremely large number of photons.

Suppose the slit is illuminated by an extremely weak light source, so weak that only 1 photon passes through the slit every second. Data are collected for 60 seconds. Draw 60 dots on the empty screen to show how you think the screen might look after 60 photons have been detected.

13. Light of wavelength $\lambda = 1\ \mu\text{m}$ is emitted from point A. A photon is detected 5 μm away at point B. On the figure, draw the trajectory that a photon follows between points A and B.

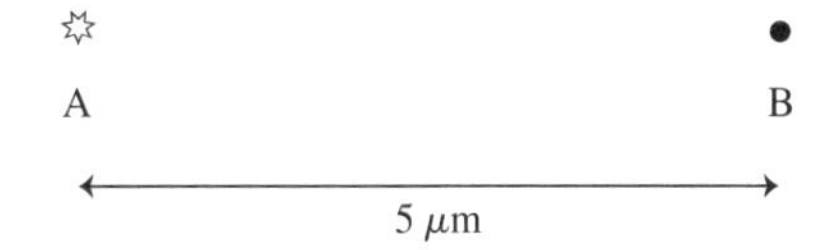

14. Three laser beams have wavelengths $\lambda_1 = 400$ nm, $\lambda_2 = 600$ nm, and $\lambda_3 = 800$ nm. The power of each laser beam is 1 W.

a. Rank in order, from largest to smallest, the photon energies E_1, E_2, and E_3 in these three laser beams.

Order:

Explanation:

b. Rank in order, from largest to smallest, the number of photons per second N_1, N_2, and N_3 delivered by the three laser beams.

Order:

Explanation:

15. When we say that a photon is a "quantum of light," what does that mean? What is quantized?

16. The intensity of a beam of light is increased but the light's frequency is unchanged. As a result:

    i. The photons travel faster.
    ii. Each photon has more energy.
    iii. The photons are larger.
    iv. There are more photons per second.

    Which of these (perhaps more than one) are true? Explain.

17. The frequency of a beam of light is increased but the light's intensity is unchanged. As a result:

    i. The photons travel faster.
    ii. Each photon has more energy.
    iii. There are fewer photons per second.
    iv. There are more photons per second.

    Which of these (perhaps more than one) are true? Explain.

28.4 Matter Waves

18. The figure is a simulation of the electrons detected behind a very narrow double slit. Each bright dot represents one electron. How will this pattern change if the following experimental conditions are changed? Possible changes you should consider include the number of dots and the spacing, width, and positions of the fringes.

    a. The electron-beam intensity is increased.

    b. The electron speed is reduced.

    c. The electrons are replaced by positrons with the same speed. Positrons are antimatter particles that are identical to electrons except that they have a positive charge.

    d. One slit is closed.

19. Very slow neutrons pass through a single, very narrow slit. Use 50 or 60 dots to show how the neutron intensity will appear on a neutron-detector screen behind the slit.

20. Electron 1 is accelerated from rest through a potential difference of 100 V. Electron 2 is accelerated from rest through a potential difference of 200 V. Afterward, which electron has the larger de Broglie wavelength? Explain.

21. An electron and a proton are each accelerated from rest through a potential difference of 100 V. Afterward, which particle has the larger de Broglie wavelength? Explain.

22. Neutron beam 1 has a temperature of 500 K. Neutron beam 2 has a temperature of 5 K. Which neutrons have the larger de Broglie wavelength? Explain.

23. A neutron is shot straight up with an initial speed of 100 m/s. As it rises, does its de Broglie wavelength increase, decrease, or not change? Explain.

24. Double-slit interference of electrons occurs because:

    i. The electrons passing through the two slits repel each other.

    ii. Electrons collide with each other behind the slits.

    iii. Electrons collide with the edges of the slits.

    iv. Each electron goes through both slits.

    v. The energy of the electrons is quantized.

    vi. Only certain wavelengths of the electrons fit through the slits.

    Which of these (perhaps more than one) are correct? Explain.

28.5 Energy Is Quantized

25. For the first allowed energy of a particle in a box to be large, should the box be very big or very small? Explain.

26. The smallest allowed energy of a particle in a box is 2.0 eV. What will the smallest energy be if:

    a. The length of the box is doubled?

    b. The mass of the particle is halved?

27. The figure shows the standing de Broglie wave of a particle in a box.

    a. What is the quantum number?

    b. Can you determine from this picture whether the "classical" particle is moving to the right or the left? If so, which is it? If not, why not?

28. A particle in a box of length L_a has $E_1 = 2$ eV. The same particle in a box of length L_b has $E_2 = 50$ eV. What is the ratio L_a/L_b?

28.6 Energy Levels and Quantum Jumps

29. The energy-level diagram is shown for a particle with quantized energy.
    a. On the left, draw an arrow or arrows to show all transitions or quantum jumps in which a particle in the $n = 3$ state absorbs a photon. Below the figure, list E_{photon} for any photon or photons that can be absorbed. Not all blanks may be needed. Don't forget units!
    b. On the right, draw an arrow or arrows to show all transitions or quantum jumps in which a particle in the $n = 3$ state emits a photon. Below the figure, list E_{photon} for any photon or photons that can be emitted. Not all blanks may be needed. Don't forget units!

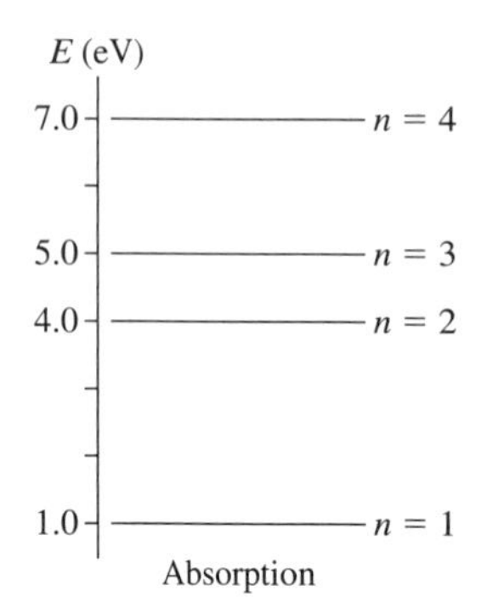

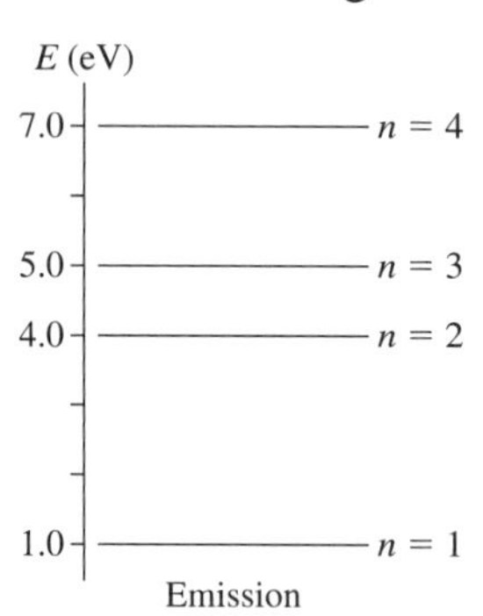

$E_{\text{photon}} =$ ______ ______ ______ $E_{\text{photon}} =$ ______ ______ ______

    c. Might this quantum system emit a photon with $E_{\text{photon}} = 3.0$ eV? If so, how could this happen? If not, why not?

28.7 The Uncertainty Principle

30. A highly collimated beam of electrons passes through a 50-nm-wide slit. An electron detector behind the slit displays a single-slit diffraction pattern with a 2.0-μm-wide central maximum. Assume that the x-axis is parallel to the plane of the slit and the plane of the detector.
    a. As an electron leaves the slit, the uncertainty in its x-position is $\Delta x =$ ______
    b. Does each electron leaving the slit have the same value of p_x? Or do different electrons have different values of p_x? Explain.
    c. Suppose the slit width is narrowed to 30 nm. Will the width of the central maximum then be greater than, less than, or equal to 2.0 μm? ______
    d. With the narrower slit, has Δx increased, decreased, or stayed the same? ______
    e. Has Δp_x increased, decreased, or stayed the same? ______

28.8 Applications and Implications of Quantum Theory

No exercises for this section.

29 Atoms and Molecules

29.1 Spectroscopy

1. The figure shows the emission spectrum of a gas discharge tube.

What color would the discharge appear to your eye? Explain.

2. A photograph of an absorption spectrum is a rainbow with black lines. A photograph of an emission spectrum is black with bright colored lines. Why are they different?

29.2 Atoms

3. Suppose you throw a small, hard rubber ball through a tree. The tree has many outer leaves, so you cannot see clearly into the tree. Most of the time your ball passes through the tree and comes out the other side with little or no deflection. On occasion, the ball emerges at a very large angle to your direction of throw. On rare occasions, it even comes straight back toward you. From these observations, what can you conclude about the structure of the tree? Be specific as to how you arrive at these conclusions *from the observations*.

4. Beryllium is the fourth element in the periodic table. Draw pictures similar to Figure 29.8 showing the structure of neutral Be, of Be^{+}, of Be^{++}, and of the negative ion Be^{-}.

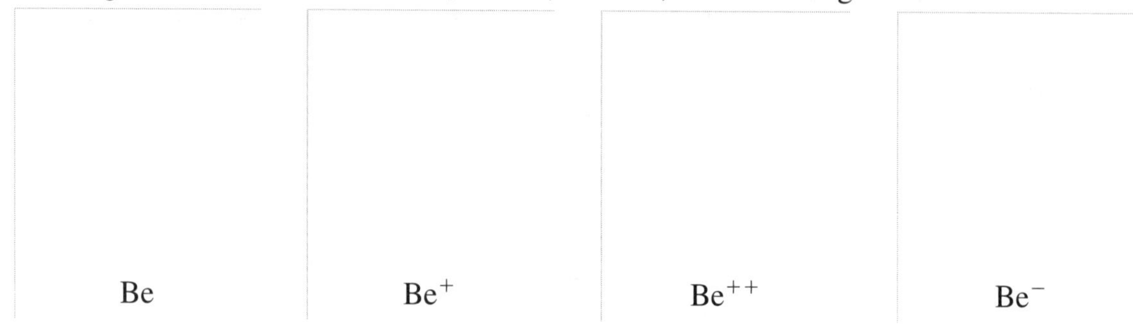

5. The element hydrogen has three isotopes. The most common has $A = 1$. A rare form of hydrogen (called *deuterium*) has $A = 2$. An unstable, radioactive form of hydrogen (called *tritium*) has $A = 3$. Draw pictures similar to Figure 29.10 showing the structure of these three isotopes. Show all the electrons, protons, and neutrons of each.

$A = 1$ $A = 2$ $A = 3$

6. Identify the element, the isotope, and the charge state. Give your answer in symbolic form, such as $^{4}He^{+}$ or $^{8}Be^{-}$.

a.

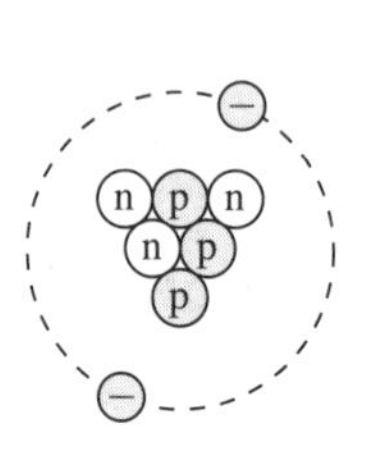

b.

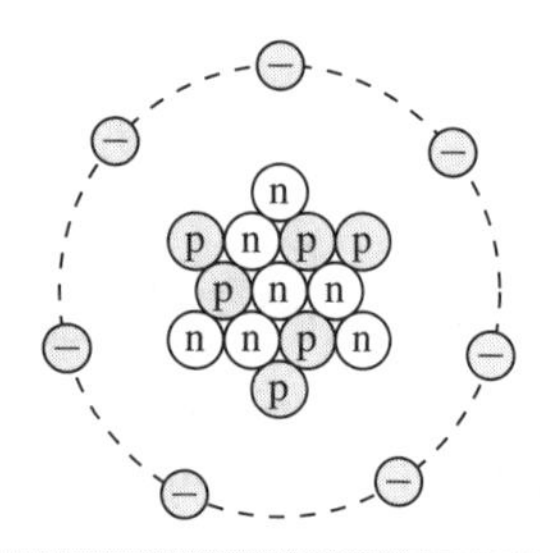

29.3 Bohr's Model of Atomic Quantization

29.4 The Bohr Hydrogen Atom

7. The figure shows a hydrogen atom, with an electron orbiting a proton.

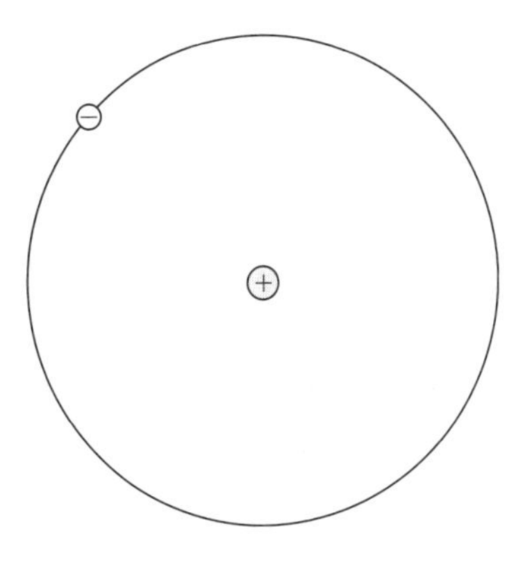

   a. What force or forces act on the electron?

   b. On the figure, draw and label the electron's velocity, acceleration, and force vectors.

8. a. The stationary state of hydrogen shown on the left has quantum number n = ________

   b. On the right, draw the stationary state of the $n - 1$ state.

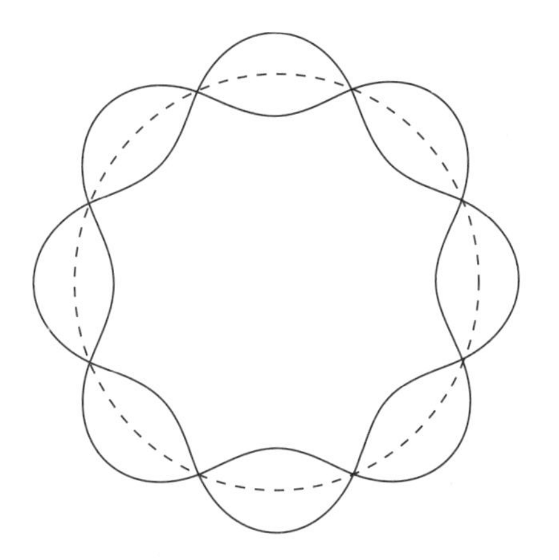

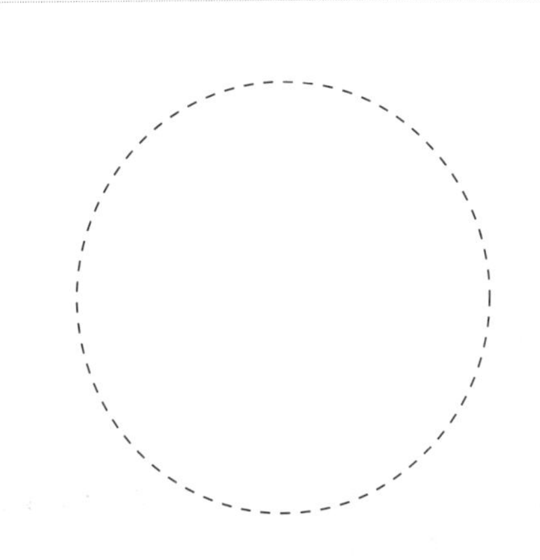

9. Bohr did not include the gravitational force in his analysis of the hydrogen atom. Is this one of the reasons that his model of the hydrogen atom had only limited success? Explain.

10. Why is there no stationary state of hydrogen with $E = -9$ eV?

11. Draw and label an energy level diagram for hydrogen. On it, show all the transitions by which an electron in the $n = 4$ state could emit a photon.

12. The longest wavelength in the Balmer series is 656 nm.
    a. What transition is this?

    b. If light of this wavelength shines on a container of hydrogen atoms, will the light be absorbed? Why or why not?

29.5 The Quantum-Mechanical Hydrogen Atom

13. List all possible states of a hydrogen atom that have $E = -1.51$ eV.

n l m

14. What are the n and l values of the following states of a hydrogen atom?

State = $4d$ $n =$ ________ $l =$ ________

State = $5f$ $n =$ ________ $l =$ ________

State = $6s$ $n =$ ________ $l =$ ________

15. How would you label the hydrogen-atom states with the following quantum numbers?

$(n, l, m) = (4, 3, 0)$ Label = ________

$(n, l, m) = (3, 2, 1)$ Label = ________

$(n, l, m) = (3, 2, -1)$ Label = ________

16. Consider the two hydrogen-atom states $5d$ and $4f$. Which has the higher energy? Explain.

29.6 Multielectron Atoms

17. Do the following figures represent a possible electron configuration of an element? If so:
    i. Identify the element, and
    ii. Determine if this is the ground state or an excited state.
    If not, why not?

a.

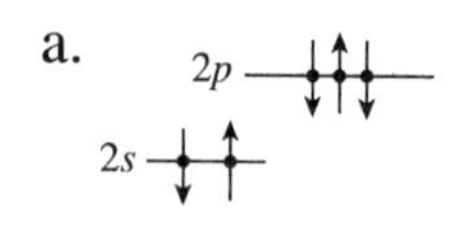

1s

b.

2p

2s

1s

c.

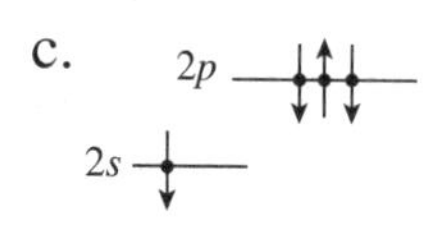

1s

18. Do the following electron configurations represent a possible state of an element? If so:
    i. Identify the element, and
    ii. Determine if this is the ground state or an excited state.
    If not, why not?

a. $1s^2 2s^2 2p^6 3s^2$

b. $1s^2 2s^2 2p^7 3s$

c. $1s^2 2s^2 2p^6 3s 3p^2$

19. Why is the section of the periodic table labeled "transition elements" exactly 10 elements wide in all rows?

29.7 Excited States and Spectra

20. The figure shows the energy levels of a hypothetical atom.

eV
0
7d
−1
7p
−2
7s
−4
6p
−7
6s
−9
Ground state

a. What is the atom's ionization energy?

b. In the space below, draw the energy-level diagram as it would appear if the ground state were chosen as the zero of energy. Label each level and the ionization limit with the appropriate energy.

21. The figure shows the energy levels of a hypothetical atom.

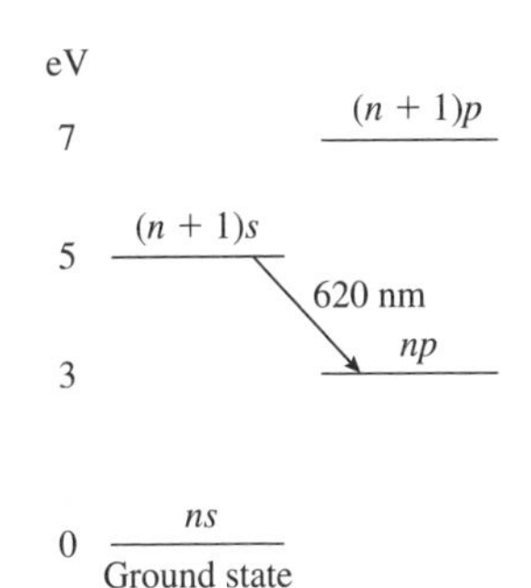

a. What *minimum* kinetic energy (in eV) must an electron have to collisionally excite this atom and cause the emission of a 620 nm photon? Explain.

b. Can an electron with $K = 6$ eV cause the emission of 620 nm light? If so, what is the final kinetic energy of the electron? If not, why not?

c. Can a 6 eV photon cause the emission of 620 nm light from this atom? Why or why not?

d. Can 7 eV photons cause the emission of 620 nm light from these atoms? Why or why not?

29.8 Molecules

22. Laundry detergent manufacturers sometimes use fluorescent chemicals to make clothes appear whiter in the presence of small amounts of UV light. What role does UV light (which is not directly visible) play in making these "whites appear whiter"?

29.9 Stimulated Emission and Lasers

23. A photon with energy 2.0 eV is an incident on an atom in the *p*-state. Does the atom undergo an absorption transition, a stimulated emission transition, or neither? Explain.

eV
3.0 *s*-state
2.0 *p*-state
0.0 *s*-state
Photon

You Write the Problem!

Exercises 24–25: You are given the equation that is used to solve a problem. For each of these:

a. Write a *realistic* physics problem for which this is the correct equation. Look at worked examples and end-of-chapter problems in the textbook to see what realistic physics problems are like.
b. Finish the solution of the problem.

24. $496 \text{ nm} = \dfrac{1240 \text{ eV} \cdot \text{nm}}{\Delta E_{\text{atom}}}$

25. $0.476 \text{ nm} = n^2 (0.0529 \text{ nm})$

$$E_n = -\frac{13.60 \text{ eV}}{n^2}$$

30 Nuclear Physics

30.1 Nuclear Structure

1. Consider the atoms ^{16}O, ^{18}O, ^{18}F, ^{18}Ne, and ^{20}Ne. Some of the questions about these atoms may have more than one answer. Give all answers that apply.

   a. Which atoms are isotopes? ________________

   b. Which atoms have the same number of nucleons? ________________

   c. Which atoms have the same chemical properties? ________________

   d. Which atoms have the same number of neutrons? ________________

   e. Which atoms have the same number of valence electrons? ________________

2. Nuclear physics calculations often involve finding the *difference* between two almost equal numbers, which requires careful attention to significant figures. Suppose you need to know the *difference* between the neutron mass and the proton mass to two significant figures. To how many significant figures must you know the masses themselves if

   a. the masses are given in atomic masses units (u)? ____________

   b. the masses are given in MeV/c^2? ____________

30.2 Nuclear Stability

3. a. Is the total binding energy of a nucleus with $A = 200$ more than, less than, or equal to the binding energy of a nucleus with $A = 60$? Explain.

   b. Is a nucleus with $A = 200$ more tightly bound, less tightly bound, or bound equally tightly as a nucleus with $A = 60$? Explain.

4. a. Is there a ^{30}Li ($Z = 3$) nucleus? If so, is it stable or radioactive? If not, why not?

b. Is there a ^{230}U ($Z = 92$) nucleus? If so, is it stable or radioactive? If not, why not?

5. Rounding slightly, the nucleus ^{3}He has a binding energy of 2.5 MeV/nucleon and the nucleus ^{6}Li has a binding energy of 5 MeV/nucleon.

a. What is the total binding energy of ^{3}He?

b. What is the total binding energy of ^{6}Li?

c. Is it energetically possible for two ^{3}He nuclei to join or fuse together into a ^{6}Li nucleus? Explain.

d. Is it energetically possible for a ^{6}Li nucleus to split or fission into two ^{3}He nuclei? Explain.

30.3 Forces and Energy in the Nucleus

6. Draw energy-level diagrams showing the nucleons in ^{6}Li and ^{7}Li.

a.

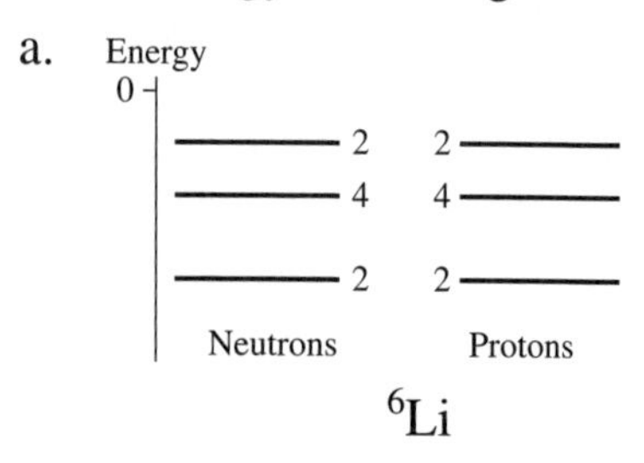

b.

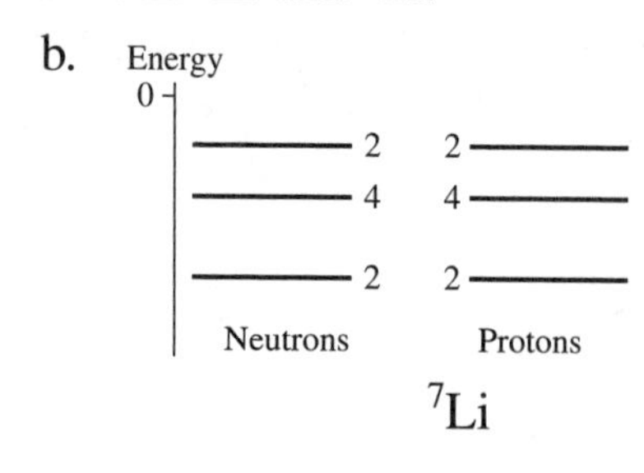

30.4 Radiation and Radioactivity

7. Identify the unknown X in the following decays:

a. ^{222}Rn (Z = 86) $\rightarrow$ ^{218}Po (Z = 84) + X   X = ____________

b. ^{228}Ra (Z = 88) $\rightarrow$ ^{228}Ac (Z = 89) + X   X = ____________

c. ^{140}Xe (Z = 54) $\rightarrow$ ^{140}Cs (Z = 55) + X   X = ____________

d. ^{64}Cu (Z = 29) $\rightarrow$ ^{64}Ni (Z = 28) + X   X = ____________

e. ^{18}F (Z = 9) + X $\rightarrow$ ^{18}O (Z = 8)   X = ____________

8. Are the following decays possible? If not, why not?

a. ^{232}Th (Z = 90) $\rightarrow$ ^{236}U (Z = 92) + α

b. ^{238}Pu (Z = 94) $\rightarrow$ ^{236}U (Z = 92) + α

c. ^{11}Na (Z = 11) $\rightarrow$ ^{11}Na (Z = 11) + γ

d. ^{33}P (Z = 15) $\rightarrow$ ^{32}S (Z = 16) + e^-

9. Part of the ^{236}U decay series is shown.

a. Complete the labeling of atomic numbers and elements on the bottom edge.

b. Label each arrow to show the type of decay.

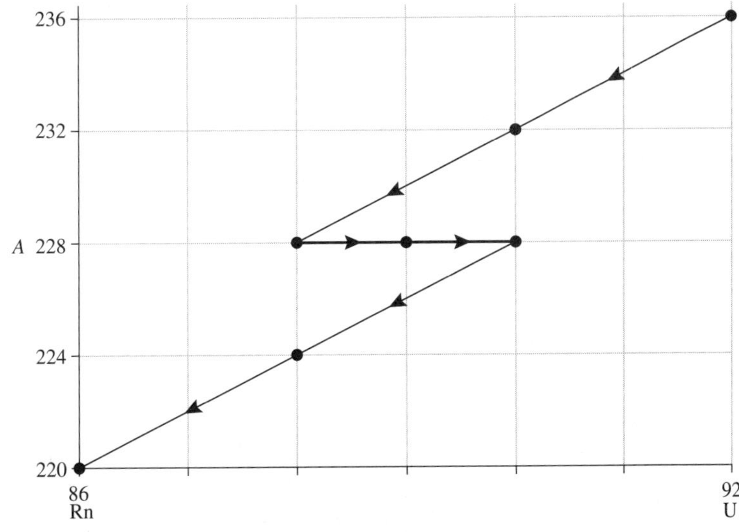

30.5 Nuclear Decay and Half-Lives

10. The half-life of the radioactive nucleus A is 6 hours.

    a. Plot the number of A nuclei remaining as a function of time on the axes below, assuming there were 1,000,000 nuclei at time $t = 0$.

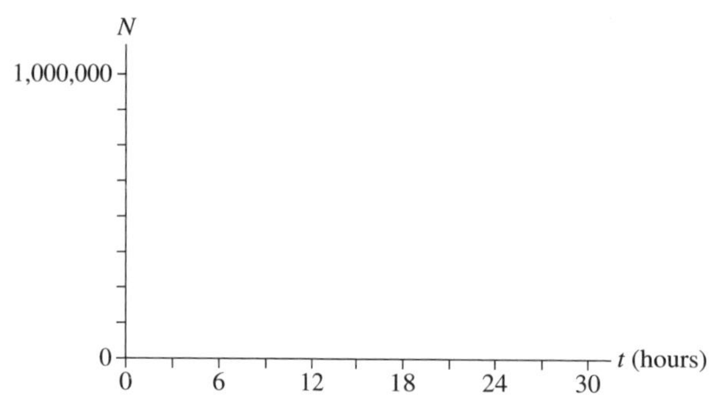

    b. How many A nuclei remain after one day?

11. What is the half-life of this nucleus?

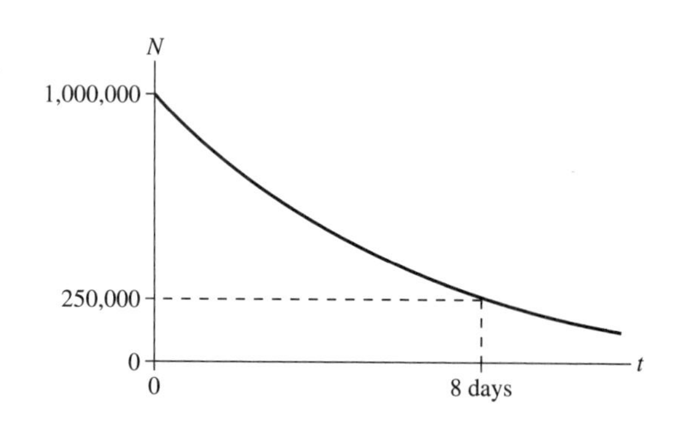

12. A radioactive sample B has a half-life of 10 s. 10,000 B nuclei are present at $t = 20$ s.

    a. How many B nuclei were there at $t = 0$ s?

    b. How many B nuclei will there be at $t = 40$ s?

13. Nucleus A decays into nucleus B with a half-life of 10 s. At $t = 0$ s, there are 1000 A nuclei and no B nuclei. At what time will there be 750 B nuclei?

30.6 Medical Applications of Nuclear Physics

14. What is the difference, or is there a difference, between radiation *dose* and *dose equivalent*?

15. An apple is irradiated for one hour by an intense beam of alpha radiation. Afterward, is the apple radioactive? Why or why not?

16. Four patients receive radiation therapy.

Patient	Radiation	Dose (Gy)	RBE
A	X rays	0.01	1
B	Alpha	0.01	20
C	Beta	0.05	1
D	Protons	0.02	5

a. Rank in order, from largest to smallest, the energy absorbed by each patient.

Order:

Explanation:

b. Rank in order, from largest to smallest, the biological damage incurred by each patient.

Order:

Explanation:

30.7 The Ultimate Building Blocks of Matter

17. A subatomic particle called the *pion* has the structure $u\bar{d}$; that is, it is made of an up quark and an antidown quark.
   a. What is the charge of a pion? Explain.

   b. What is the structure of an antipion? What is its charge?

18. One family of subatomic particles made of three up (u) and/or down (d) quarks is the Δ family. The Δ^{++} particle has the interesting property that its charge is $+2e$. What is the quark structure of the Δ^{++}? Explain.

You Write the Problem!

Exercises 19–20: You are given the equation that is used to solve a problem. For each of these:

a. Write a *realistic* physics problem for which this is the correct equation. Look at worked examples and end-of-chapter problems in the textbook to see what realistic physics problems are like.

b. Finish the solution of the problem.

19. $3.75 \times 10^{12} = 6.00 \times 10^{13}\left(\frac{1}{2}\right)^{(2.5\text{ min})/t_{1/2}}$

20. $3.0\text{ Sv} = \frac{\text{dose}}{0.035\text{ kg}} \times 5$